A Choice of
Catastrophes

ABOUT THE AUTHOR

Isaac Asimov has published more than 200 books, most of them on scientific subjects. His output includes superb explanatory works on astronomy, biology, and chemistry, as well as a biographical encyclopedia of science. And, of course, he has been a science-fiction pioneer, producing some of that genre's most celebrated works. In the words of Dr. Carl Sagan, Mr. Asimov "is a natural resource, a Renaissance man born out of his time—thank God."

Also by Isaac Asimov:

EARTH: OUR CROWDED SPACESHIP

EXTRATERRESTRIAL CIVILIZATIONS

ISAAC ASIMOV'S BOOK OF FACTS

REALM OF ALGEBRA

REALM OF NUMBERS

A Choice of

The Disasters That Threaten

Catastrophes

Our World *by Isaac Asimov*

FAWCETT COLUMBINE • NEW YORK

A CHOICE OF CATASTROPHES

Published by Fawcett Columbine Books, a unit of CBS Publications, the Consumer Publishing division of CBS Inc., by arrangement with Simon and Schuster, a division of Gulf & Western Corporation.

ISBN: 0-449-90048-7

Printed in the United States of America

First Fawcett Columbine Printing: April 1981

10   9   8   7   6   5   4   3   2

To Robyn and Bill,
may they find Fortune's face
a smiling one always

Contents

Introduction

The word "catastrophe" is from the Greek and it means "to turn upside down." It was originally used to describe the denouement, or climactic end, of a dramatic presentation and it could, of course, be either happy or sad in nature.

In a comedy, the climax is the happy ending. After a spate of misunderstandings and sorrow, everything is turned upside down when the lovers are suddenly reconciled and reunited. The catastrophe of the comedy is, then, an embrace or a marriage. In a tragedy, the climax is the sad ending. After endless striving, everything is turned upside down when the hero finds that fate and circumstance defeat him. The catastrophe of the tragedy is, then, the death of the hero.

Since tragedies tend to strike deeper than comedies do and to be more memorable, the word "catastrophe" has come to be associated more with tragic endings than with happy ones. Consequently, it's now used to describe any final end of a calamitous or disastrous nature—and it is with that kind of catastrophe that this book deals.

The final end of what? Of ourselves, of course; of the human species. If we regard human history as a tragic drama, then the final death of humanity would be the catastrophe in both the original and the present sense. But what could bring about the end of human history?

For one thing, the entire universe might so change its properties as to become uninhabitable. If the universe became deadly and if no life could exist anywhere within it, then humanity could not exist either, and that would be something we might call a "Catastrophe of the First Class."

Naturally, the entire universe need not be involved in something that would suffice to bring about an end to humanity. The universe might be as benign as it now is and yet something might happen to the sun that would make the solar system uninhabitable. In that case, human life might come to an end even though all the rest of the universe might be proceeding on its way smoothly and peacefully. This would be a "Catastrophe of the Second Class."

To be sure, though the sun might continue to shine as evenly and as benevolently as ever, the Earth itself might undergo the kind of convulsion that would make life impossible upon it. In that case, human life might come to an end even though the solar system continued on its routine round of rotations and revolutions. This would be a "Catastrophe of the Third Class."

And though Earth might remain warm and pleasant, something might happen upon it that would destroy human life, while leaving at least some other forms of life untouched. In that case, evolution might continue and Earth, with a modified load of life, might flourish —but without us. That would be a "Catastrophe of the Fourth Class."

We might even go one step farther and point out the possibility that human life might continue, but that something would happen that would destroy civilization, interrupting the march of technological advance and condemning humanity to a primitive life—solitary, poor, nasty, brutish, and short—for an indefinite period. This would be a "Catastrophe of the Fifth Class."

In this book, I will take up all these varieties of catastrophes beginning with the first class and following with the others in order. The catastrophes described will be successively less cosmic—and successively more immediate and dangerous.

Nor need the picture so drawn be one of unalloyed gloom, since it may well be that there is no catastrophe that is not unavoidable. And certainly the chances of avoiding one increase, if we stare the catastrophe boldly in the face and estimate its dangers.

Part I

Catastrophes of the First Class

Chapter 1

The Day of Judgment

Ragnarok

The conviction that the whole universe is coming to an end (the catastrophe of the first class mentioned in the introduction) is an old one, and is, in fact, an important part of Western tradition. One particularly dramatic picture of the end of the world is given to us of the Western tradition in the myths that originated among the Scandinavian peoples.

The Scandinavian mythology is a reflection of the harsh, subpolar environment in which the hardy Norsemen lived. It is a world in which men and women play a minor part, and in which the drama rests in the conflict between gods and giants, a conflict in which the gods seem at a perpetual disadvantage.

The frost-giants (the long, cruel Scandinavian winters) are undefeatable, after all, and even within the beleaguered fortress of the gods themselves, Loki (the god of the fire that is so essential in a northern climate) is as tricky and as treacherous as fire itself is. And in the end there comes *Ragnarok,* which means "the fatal destiny of the gods." (This term Richard Wagner made better known as *Götterdämmerung,* or "twilight of the gods," in his opera of that name.)

Ragnarok is the final decisive battle of the gods and their enemies. Behind the gods come the heroes of Valhalla who, on Earth, had died in battle. Opposed are the giants and monsters of a cruel nature led by the renegade, Loki. One by one the gods fall, though the monsters and giants—and Loki, too—also die. In the fight, the Earth and the universe perish. The sun and the moon are swallowed by the wolves who have been pursuing them since creation. The Earth catches fire and bakes and cracks in a universal holocaust. Almost as an insignificant side issue to the great battle, life and mankind are wiped out.

And that should, dramatically, be the end—but it isn't.

Somehow a second generation of gods survives; another sun and moon come into being; a new Earth arises; a new human pair comes into existence. An anticlimactic happy ending is tacked on to the grand tragedy of destruction. How did this come to be?

The tale of Ragnarok, as we now have it, is taken from the writings of an Icelandic historian, Snorri Sturluson (1179–1241). By that time, Iceland had been Christianized and the tale of the end of the gods seems to have suffered a strong Christian influence. There were, after all, Christian tales of the death and regeneration of the universe that long antedated the Icelandic tale of Ragnarok, and the Christian tales were, in turn, influenced by those of the Jews.

Messianic Expectations

While the Davidic kingdom of Judah existed, prior to 586 B.C., the Jews were quite certain God was the divine judge who meted out rewards and punishments to individuals in accordance with their deserts. The rewards and punishments were handed out in this life. This confidence did not survive defeat.

After Judah had been cast down by the Chaldeans under Nebuchadnezzar, after the Temple had been destroyed and many of the Jews brought to exile in Babylonia, there grew up a longing among the exiles for the return of the kingdom and of a king of the old Davidic dynasty. Since such desires, too plainly expressed, represented treason to the new non-Judaic rulers, the habit arose of speaking of the return of the king elliptically. One spoke of "the Messiah";

that is, "the anointed one" since the king was anointed with oil as part of the ritual of assuming office.

The picture of the returning king was idealized as introducing a wonderful golden age and, indeed, the rewards of virtue were removed from the present (where it manifestly was not taking place) and put into a golden future.

Some verses describing that golden age were placed in the Book of Isaiah, which dealt with the words of a prophet who preached as early as 740 B.C. The verses themselves probably came from a later period. Of course, in order to introduce the golden age, the righteous among the population must be advanced to power, and the evildoers must be rendered powerless or even destroyed. Thus:

"And he [God] shall judge among the nations, and shall rebuke many people: and they [the nations] shall beat their swords into plowshares and their spears into pruninghooks: nation shall not lift up sword against nation, neither shall they learn war any more" (Isa. 2:4).

". . . with righteousness shall he [God] judge the poor, and reprove with equity for the meek of the earth: and he shall smite the earth with the rod of his mouth, and with the breath of his lips shall he slay the wicked" (Isa. 11:4).

Time passed and the Jews returned from exile, but that brought no relief. There was hostility from their immediate non-Jewish neighbors and they felt helpless against the overwhelming power of the Persians, who now ruled over the land. The Jewish prophets grew all the more graphic, therefore, in their writings about the coming golden age, and particularly about the doom awaiting their enemies.

The prophet, Joel, writing about 400 B.C., said, "Alas for the day! for the day of the Lord is at hand, and as a destruction from the Almighty shall it come" (Joel 1:15). The picture is of the coming of a specific time when God shall judge all the world: "I will also gather all nations, and will bring them down into the valley of Jehoshaphat, and will plead with them there for my people and for my heritage Israel . . ." (Joel 3:2). This was the first literary expression of a "Judgment Day," a time when God would bring to an end the present order of the world.

The notion became stronger and more extreme in the second century B.C., when the Seleucids, the Greek rulers who had succeeded to the Persian dominion after the time of Alexander the Great, tried to suppress Judaism. The Jews, under the Maccabees, rebelled, and the Book of Daniel was written to support the rebellion and to promise a glowing future.

The book drew, in part, on older traditions concerning a prophet, Daniel. Into Daniel's mouth were placed descriptions of apocalyptic visions.* God (referred to as "the Ancient of days") makes his appearance to punish the wicked:

"I saw in the night visions, and, behold, one like the Son of man came with the clouds of heaven, and came to the Ancient of days, and they brought him near before him. And there was given him dominion, and glory, and a kingdom, that all people, nations, and languages, should serve him: his dominion is an everlasting dominion, which shall not pass away, and his kingdom that which shall not be destroyed" (Dan. 7:13–14).

This "one like the Son of man" refers to someone in human shape as contrasted to the enemies of Judah who had previously been portrayed in the shape of various beasts. The human shape can be interpreted as representing Judah in the abstract, or the Messiah in particular.

The Maccabean rebellion was successful and a Judean kingdom was reestablished but this did not bring the golden age, either. However, the prophetic writings kept expectations keen among the Jews over the next couple of centuries. The Day of Judgment remained always about to come; the Messiah was always at hand; the kingdom of righteousness was always on the point of being established.

The Romans took over from the Maccabees and in the reign of the emperor Tiberius, there was a very popular preacher in Judea named John the Baptist, and the burden of his message was "Repent ye: for the kingdom of heaven is at hand" (Matt. 3:2).

With universal expectation thus constantly sharpened, anyone who claimed to be the Messiah was bound to raise a following, and under the Romans there were a number of such claimants who came to nothing, politically. Among such claimants, however, was Jesus of Nazareth, whom a few humble Judeans followed, and who remained faithful even after Jesus had been crucified without a hand lifted to save him. Those who believed in Jesus as the Messiah might have been called "Messianics." However, the language of Jesus' followers came to be Greek as more and more Gentiles were converted, and in Greek the word for Messiah is "*Christos.*" Jesus' followers came to be called "Christians."

---

* "Apocalyptic" is from Greek words meaning "disclosing" so anything which is apocalyptic discloses a future ordinarily hidden from human eyes.

The early success in converting Gentiles came through the charismatic missionary preaching of Saul of Tarsus (the Apostle Paul) and beginning with him, Christianity began a career of growth that brought first Rome, then Europe, then much of the world to its banner.

The early Christians believed that the arrival of Jesus the Messiah (that is "Jesus Christ") meant that the Day of Judgment was at hand. Jesus himself was described as making predictions of an imminent end of the world:

"But in those days, after that tribulation, the sun shall be darkened, and the moon shall not give her light, and the stars of heaven shall fall, and the powers that are in heaven shall be shaken. And then shall they see the Son of man coming in the clouds with great power and glory. . . . Verily I say unto you, that this generation shall not pass, till all these things be done. Heaven and earth shall pass away. . . . But of that day and that hour knoweth no man, no, not the angels which are in heaven, neither the Son, but the Father" (Mark 13:24–27, 30–32).

About 50 A.D., twenty years after the death of Jesus, the Apostle Paul still expected the Day of Judgment momentarily:

"For this we say unto you by the word of the Lord, that we which are alive and remain unto the coming of the Lord shall not prevent them which are asleep. For the Lord himself shall descend from heaven with a shout, with the voice of the archangel, and with the trump of God: and the dead in Christ shall rise first: Then we which are alive and remain shall be caught up together with them in the clouds, to meet the Lord in the air: and so shall we ever be with the Lord. Wherefore comfort one another with these words. But of the times and the seasons, brethren, ye have no need that I write unto you. For yourselves know perfectly that the day of the Lord so cometh as a thief in the night" (1 Thess. 4:15, 5:2).

Paul, like Jesus, implied that the Day of Judgment would come soon but was careful not to set an exact date. And, as it happened, the Day of Judgment did *not* come; the evil were *not* punished, the ideal kingdom was *not* set up, and those who believed that Jesus was the Messiah had to content themselves with the feeling that the Messiah would have to come a second time (the "Second Coming") and that *then* all that had been foretold would come to pass.

The Christians were persecuted in Rome under Nero, and on a wider scale under the later emperor Domitian. Just as the Seleucid persecution had brought forth the apocalyptic promises of the Book of Daniel in Old Testament times, so the persecutions of Domitian brought forth the apocalyptic promises of the Book of Revelation in New Testament times. Revelation was probably written in 95 A.D. during the reign of Domitian.

In great, and utterly confusing, detail, the Day of Judgment is pictured. There is talk of a final battle between all the forces of evil and the forces of good at a place called Armageddon, though the details aren't clear (Rev. 16:14–16). Finally, though, "I saw a new heaven and a new earth: for the first heaven and the first earth were passed away . . ." (Rev. 21:1).

It is quite possible, then, that whatever the Scandinavian myth of Ragnorak may have been to begin with, the version that has come to us must owe something to that battle of Armageddon in Revelation with its vision of a regenerated universe. And Revelation in turn owes a great deal to the Book of Daniel.

Millennarianism

The Book of Revelation introduced something new: "And I saw an angel come down from heaven, having the key of the bottomless pit and a great chain in his hand. And he laid hold on the dragon, that old serpent, which is the Devil, and Satan, and bound him a thousand years. And cast him into the bottomless pit, and shut him up, and set a seal upon him, that he should deceive the nations no more, till the thousand years should be fulfilled: and after that he must be loosed a little season" (Rev. 20:1–3).

Why the devil is to be put out of action for a thousand years or "millennium" and then turned loose "a little season" is not clear, but at least it lifted the pressure from those who believed that the day of judgment was at hand. One could always say that the Messiah had come and that the devil was in bonds, meaning that Christianity could give strength—but the true final battle and the true end would come a thousand years later.*

It seemed natural to suppose that the thousand years had begun ticking away with the birth of Jesus and in the year 1000 there was a flurry of nervous apprehension, but it passed—and the world did not pass.

---

* In fact, it is because of the thousand-year-binding of Satan that the term "millennium" has come to be applied to a period of future ideal justice and happiness, often used ironically as something that would never happen.

The words of Daniel and of Revelation were so elliptical and obscure, and the urge to believe was so great, however, that it always remained possible for people to reread those books, reweigh the vague predictions, and come up with new dates for the Day of Judgment. Even great scientists, such as Isaac Newton and John Napier played that game.

Those who tried to calculate when that crucial thousand years would start and end are sometimes called "millennialists" or "millennarians." They can also be called "chiliasts" from the Greek word for a thousand years. Oddly enough, Millennarianism, despite repeated disappointments, is stronger now than ever.

The current movement began with William Miller (1782–1849), an army officer who fought in the War of 1812. He had been a skeptic but after the war he became what we would now call a born-again Christian. He began to study Daniel and Revelation and decided that the Second Coming would take place on March 21, 1844. He supported it by involved calculations and predicted that the world would end in fire after the fashion of the lurid descriptions of the Book of Revelation.

He gained a following of as many as 100,000 people, and on the day appointed, many of them, having sold their worldly goods, gathered on hillsides to be swept upward to meet Christ. The day passed without incident, whereupon Miller recalculated the matter and set October 22, 1844, as the new day, and that passed without incident also. When Miller died in 1849, the universe was still in business.

Many of his followers were not discouraged, however. They interpreted the apocalyptic books of the Bible in such a way as to have Miller's calculations indicate the beginning of some heavenly process as yet invisible to the ordinary consciousness on Earth. It was still another "millennium" of waiting, after a fashion, and the actual Second Coming, or "Advent" of Jesus was postponed once again into the future—but, as always before, the not-too-distant future.

Thus was founded the Adventist movement, which split up into a number of different sects, including the Seventh-Day Adventists who returned to such Old Testament observances as keeping the Sabbath on Saturday (the seventh day).

One person who adopted Adventist views was Charles Taze Russell (1852–1916) who, in 1879, founded an organization that came to be called Jehovah's Witnesses. Russell expected the Second Coming momentarily and predicted it on several different days after the fashion of Miller, each time being disappointed. He died during World

War I, which must have seemed to him like the opening of the final, climactic battles described in Revelation—but still the Advent did not follow.

The movement continued to flourish, however, under Joseph Franklin Rutherford (1869–1942). He awaited the Second Coming with the stirring slogan, "Millions now living shall never die." He himself died during World War II, which again must have seemed like the opening of the final, climactic battles described in Revelation—and still the Advent did not follow.

But the movement flourishes anyway and now claims a world membership of over a million.

Chapter 2

The Increase
of Entropy

The Conservation Laws

So much for the "mythic universe." Along with the mythical outlook, however, there has been a scientific view of the universe, one which deals with observation and experiment (and, occasionally, intuitive insights which must then, however, be backed by observation and experiment).

Suppose we consider this scientific universe (as we shall in the remainder of the book). Is the scientific universe, like the mythic universe, fated to come to an end? If so, how, and why, and when?

The ancient Greek philosophers felt that while Earth was the home of change, corruption and decay, the heavenly bodies followed different rules and were changeless, incorruptible, and eternal. The medieval Christians felt that the sun, moon, and stars would meet the common ruin of the Day of Judgment but till then they were, if not eternal, at least changeless and incorruptible.

The view began to change when the Polish astronomer Nicolas Copernicus (1473–1543) published a carefully reasoned book in 1543, one in which Earth was removed from its unique position at the center of the universe, and was viewed as a planet which, like other planets, circled the sun. It was the sun that now took over the unique central position.

Naturally, the Copernican view was not adopted immediately and, in fact, was violently opposed for sixty years. It was the coming of the telescope, first used to view the sky in 1609 by the Italian scientist Galileo (1564–1642), that removed opposition from any claim to scientific respectability and reduced it to mere stubborn obscurantism.

Galileo discovered, for instance, that Jupiter had four satellites that circled it steadily thus disproving once and for all that Earth was the center about which *all* things turned. He found that Venus showed a full cycle of moonlike phases, as Copernicus predicted it would have to, where earlier views had predicted otherwise.

Through his telescope, Galileo also saw the moon to be covered by mountains, craters, and what he took to be seas, showing that it (and by extension the other planets) were worlds like the Earth and, therefore, presumably subject to the same laws of change, corruption, and decay. He detected dark spots on the surface of the sun itself, so that even this transcendent object, which, of all material things, seemed the closest approach to the perfection of God, was, after all, imperfect.

In the search for the eternal, then—or at least for those aspects of the eternal that could be observed and were therefore part of the scientific universe—people had to reach for a more abstract level of experience. If it was not things that were eternal, perhaps it was relationships among things.

In 1668, for instance, the English mathematician John Wallis (1616–1703) investigated the behavior of colliding bodies and came up with the notion that in the process of collision, some aspect of movement doesn't change.

Here's the way it works. Every moving body has something called "momentum" (which is the Latin word for "movement"). Its momentum is equal to its mass (which may be roughly defined as the amount of matter it contains) multiplied by its velocity. If the movement is in one particular direction, the momentum can be given a positive sign; in the opposite direction a negative sign.

If two bodies approach each other head-on, there will be a total momentum which we can determine by subtracting the minus-momentum of one from the plus-momentum of the other. After they strike each other and recoil, the distribution of momentum between the two bodies will change, but the total momentum will be the same as before. If they collide and stick, the new combined body will have a different mass from either separately and a different velocity from either, but the total momentum will stay the same. The total momentum stays the same even if the bodies hit at an angle instead of head-on and bounce away in changed directions.

From Wallis's experiments and from many others made since, it turns out that in any "closed system" (one in which no momentum enters from the outside and no momentum vanishes into the outside) the total momentum always remains the same. The distribution of the momentum among the moving bodies in the system may change in any of an infinite number of ways, but the total remains the same. Momentum is therefore "conserved"; that is, it is neither gained nor lost; and the principle is called "the law of conservation of momentum."

Since the only truly closed system is the whole universe, the most general way of stating the law of conservation of momentum is to say "the total momentum of the universe is constant." In essence, it never changes through all eternity. No matter what changes have taken place, or may yet take place, the total momentum does not change.

How can we be sure? How can we tell from a few observations made by scientists under laboratory conditions over a few centuries that momentum will be conserved a million years from now, or was conserved a million years ago? How can we tell whether it is conserved right now a million light-years away in another galaxy, or right in our neighborhood under conditions as alien as those in the center of the sun?

We *can't* tell. All we can say is that at no time under any conditions have we observed the law violated; nor have we detected anything which indicates that it ever might be violated. Furthermore, all the consequences we deduce on the assumption that the law is true seem to make sense and to fit in with what is observed. Scientists therefore feel they have ample right to *assume* (always pending evidence to the contrary) that the conservation of momentum is a "law of nature" that holds universally through all of space and time and under all conditions.

The conservation of momentum was only the first of a series of conservation laws worked out by scientists. For instance, one can speak of "angular momentum," which is a property possessed by bodies that turn around an axis of rotation, or around a second body elsewhere. In either case, one calculates angular momentum from the mass of a body, its velocity of turning, and the average distance of its parts from the axis or center about which the turning takes place. It turns out there is a law of conservation of angular momentum. The total angular momentum of the universe is always constant.

What's more, the two types of momentum are independent of each other and are not interchangeable. You can't change angular momentum into ordinary momentum (sometimes called "linear momentum" to differentiate it from the other) or vice versa.

In 1774, a series of experiments by the French chemist Antoine-Laurent Lavoisier (1743–94) suggested that mass was conserved. Within a closed system, some bodies might lose mass and others might gain it, but the total mass of the system remained constant.

Gradually, the scientific world developed the concept of "energy," that property of a body which enables it to do work. (The very word, energy, is from a Greek expression meaning "containing work.") The first to use the word in its modern sense was the English physicist Thomas Young (1773–1829), in 1807. A variety of different phenomena were all capable of doing work—heat, motion, light, sound, electricity, magnetism, chemical change, and so on—and all came to be considered different forms of energy.

The notion grew that one form of energy could be converted into another, that some bodies might lose energy in one form or another and that other bodies might gain energy in one form or another, but that in any closed system, the total energy of all forms was constant. By no means the first to think so was the German physicist Hermann L. F. von Helmholtz (1821–94), but he managed, in 1847, to persuade the scientific world generally that this is so. He is usually considered, therefore, the discoverer of the law of conservation of energy.

In 1905, the German-Swiss physicist Albert Einstein (1879–1955) was able to argue convincingly that mass was one more form of energy, that a given quantity of mass could be converted into a fixed quantity of energy, and vice versa.

For that reason the law of conservation of mass disappeared as a separate conservation law, and one speaks only of the law of conservation of energy these days, it being understood that mass is included as a form of energy.

Once the structure of the atom was determined by the British physicist Ernest Rutherford (1871–1937) in 1911, it was found that there

existed subatomic particles, which not only followed the laws of conservation of momentum, angular momentum, and energy, but also
the laws of conservation of electric charge, baryon number, isotopic
spin, and a few other such rules.

The various conservation laws are, indeed, the basic rules of the
game played by all the bits and pieces of the universe; and all those
laws are general and eternal as far as we know. If a conservation law
turns out not to be valid after all, then this proves to be so because it
is part of a more general law. Thus, the conservation of mass turned
out to be invalid, but part of a more general conservation of energy
that includes mass.

Now we have one aspect of the universe that would seem to be
eternal and with neither beginning nor end. The energy the universe
now contains will always be there in precisely the same quantity as
now and has always been there in precisely the same quantity as
now. Ditto the momentum, the angular momentum, the electric
charge, and so on. There will be all sorts of local changes as this part
or that part of the universe loses or gains one of these conserved
properties, or has one of the conserved properties change its form—
but the total was, is, and will be, unchanged.

Energy Flow

We can now draw a parallel between the mythic universe and the
scientific universe.

In the case of the mythic universe, there is an eternal and undecaying heavenly kingdom against which is the changing world of the flesh
with which we are familiar. It is this changing world which we think
of as coming to an end; it is only this changing world concerning
which the word "end"—or "beginning," for that matter—has meaning. It is not only changing; it is temporary.

In the scientific universe, there are the eternal and undecaying
conserved properties against which is a changing world that plays
itself out against the background of and according to the rules of
those conserved properties. It is only this changing world concerning
which the word "end"—or "beginning"—has meaning. It is not only
changing, but it is temporary.

But why should there be a changing and temporary aspect of the
scientific universe? Why don't all the components of the universe get
together into one super-massive object with some certain momentum,

angular momentum, electric charge, energy content, and so on, and then never change?

Why, instead, does the universe consist of a myriad of objects of all sizes that constantly transfer bits of the conserved properties from one to another?*

The driving power behind all these changes is, apparently, energy, so that, in a way, energy is the most important property the universe possesses, and the law of conservation of energy is considered by some to be the most basic of all the laws of nature.

Energy drives all the changes in the universe by itself participating in changes. Bits of energy flow from one place to another, from one body to another, changing in form at times as they do so. This means we have to ask what it is that drives the energy this way and that.

The reason for this, apparently, is that energy is spread through the Universe in uneven fashion; it is present in more concentrated form in some places and in less concentrated form in other places. All the flow of bits of energy from one place to another, from one body to another, from one form to another, takes place in such a way that the tendency is to even out the distribution.† It is the energy flow, that converts an uneven distribution to an even one, that can be used to do work and to bring about all the changes we see taking place; all the changes we associate with the universe as we know it, with life and with intelligence.

What's more, the evening-out of energy is spontaneous. Nothing has to drive the energy flow necessary to bring it about. It takes place by itself. It is self-driving.

Let me give you a simple example. Suppose you have two large containers of equal size connected near the bottoms by a horizontal tube which is blocked so that no actual communication exists between the two containers. You can fill one of the containers with water all the way up to the top, while in the other you can put only a little bit of water.

The container that is full has its water higher, on the average, than the container that is nearly empty. To lift water higher against the pull of gravity requires an input of energy so that the water in the full container has a higher level of energy with respect to the gravitational

* We can't object to this, of course, since it is the constant transfer back and forth of conserved properties that produces all the activity, animate and inanimate, in the universe; that makes life possible; that produces the restless evanescence we call intelligence, and so on.
† Of course, we then have to ask why the energy is unevenly distributed in the first place. We'll take up that question later.

field than the water in the nearly empty container. For historical reasons, we say that the water in the full container has more "potential energy" than the water in the nearly empty container.

Imagine, now, that the tube connecting the two containers is opened. Promptly, water will flow from the place where it contains a higher potential energy to the place where it contains a lower one. Water will flow from the full container to the nearly empty one—spontaneously.

There's no question in anyone's mind, I'm sure, provided that that mind has had the least experience with the world, that this is a spontaneous and unavoidable event. If the tube were opened and the water failed to flow from the full container to the nearly empty one, we would decide at once that the connecting tube was not open after all but was still blocked. If what little water was in the nearly empty container were to flow into the full container, we would have to decide the water was being pumped.

If the tube were undeniably open and if it were clear that no pumping was involved and if the water did not flow from the full container to the nearly empty one, or if, worse still, the water flowed in the other direction, we would have to come to the worried conclusion that we were witnessing what could only be described as a miracle. (Needless to say, no such miracle has ever been witnessed and recorded in the annals of science.*)

In fact, so certain is the spontaneous flow of water in this fashion that we use it, automatically, as a measure of the direction of time-flow.

Suppose, for instance, that someone had taken a motion picture of events in the two containers, and we were watching the results. The connecting tube is opened and yet the water doesn't flow. We would at once come to the conclusion that the film wasn't running and that we were watching a "still." In the movie universe, in other words, time had come to a halt.

Again, suppose that the movie showed us water flowing from the nearly empty container into the full container. We would be quite certain that the film was running backward. In the movie universe, the direction of time-flow was the reverse of what it is in real life. (In fact, the effect of running a movie film backward is almost invariably humorous because there are innumerable events that then happen that we know never happen in real life. Splashing water draws itself

* The parting of the Red Sea as portrayed in the motion picture *The Ten Commandments* is precisely such a miracle, by the way. Naturally, it requires trick photography.

inward while a diver heaves out of the water feet-first and lands on a diving board; the fractured shards of a glass draw themselves together and fit themselves perfectly into an intact object; wind-blown hair is wafted into a perfect coiffure. Watching any of this makes us realize how many events in real life are clearly spontaneous; how many reversals, if they actually took place, would seem clearly miraculous; and how well we know one from the other simply through experience.)

Returning to the two containers of water, it is easy to show that the rate at which the water flows from the full container to the nearly empty one depends on the difference in the energy distribution. At the start, the potential energy of the water in the full container is considerably greater than the potential energy of the water in the nearly empty one, so the water flows quickly.

As the water level drops in the full container and rises in the empty one, the difference in potential energy between the two containers decreases steadily, so that the distribution of energy is less uneven, and the water flows at a steadily decreasing rate. By the time the levels of water are almost even, the water is flowing at a very slow rate and when the levels of water in the two containers are quite even, and there is no potential energy difference at all between them, the water flow stops altogether.

In short, the spontaneous change is from a state of uneven distribution of energy to a state of even distribution of energy, and at a rate that is proportional to the amount of unevenness. Once the even distribution of energy is achieved, change *stops*.

If we were to watch two connected containers of water, with the water level equal in both, and with no intervention from the outside at all, then if water flowed in either direction so that the level in one rose and the level in the other dropped, we would be witnessing a miracle.

The moving water can do work. It can turn a turbine which will generate a flow of electricity, or it can simply push things along with it. As the rate of water flow slows, the rate at which work can be done slows with it. When the water flow stops altogether, no further work can be done.

When the water flow stops, when the height of water is the same in both containers, then everything stops. All the water is still there. All the energy is still there. All that water and energy, however, *is no longer unevenly distributed*. It is the uneven distribution of energy that produces change, motion, work, as it strives toward even distribution. Once the even distribution is achieved, there is, thereafter, no change, no motion, no work.

What's more, the spontaneous change is always from uneven distribution to even distribution, and once the even distribution is reached, nothing spontaneous will ever change it back to an uneven distribution.*

Let's take another example; one that involves heat rather than water level. Of two bodies, one may contain a higher intensity of heat energy than the other. The level of intensity of heat energy is measured as "temperature." The higher the level of intensity of heat energy of a body, the higher its temperature and the hotter it is. We can therefore speak of a hot body and a cold body and find them equivalent to our earlier case of the full container and the nearly empty container.

Suppose that the two bodies formed a closed system so that no heat could flow into them from the outside universe and no heat could flow out of them into the outside universe. Now imagine the two bodies, the hot one and the cold one, brought into contact.

We know exactly what would happen from our experience with real life. Heat will flow from the hot body into the cold body, just as water will flow from a full container into an empty one. As the flow of heat continues, the hot body will cool down and the cold body will warm up, just as the full container grew less full and the empty container grew fuller. Finally, the two bodies will be at the same temperature, just as the two containers ended with the same water level.

Again, the rate of heat flow from the hot body to the cold body depends on the amount of unevenness of energy distribution. The greater the difference in temperature between the two bodies the more rapidly heat will flow from the hot body to the cold one. As the hot body cools and the cold body warms, the temperature difference decreases and so does the rate of flow of heat. Finally, when the two bodies are at the same temperature, the flow of heat stops altogether and moves in neither direction.

Again, this direction of heat flow is spontaneous. If two bodies of different temperatures were brought together and if heat did not flow, or if heat flowed from the cold body into the hot body so that the cold body grew still colder and the hot body still hotter—and if we were sure we were dealing with a really closed system and there was no

---

* Actually, as we shall see, this is not *quite* true.

hanky-panky—then we would have to conclude we were witnessing a miracle. (And again no such miracle has been witnessed and recorded by scientists.)

Then, too, once the two bodies are at the same temperature, any heat flow that would cause either of the two bodies to grow warmer or cooler does not take place.

Such changes are once again related to the flow of time. If we took a movie of the two objects, focusing on a thermometer attached to each, and noticed that one temperature remained high and one low, with no change, we would conclude that the film was not moving. If we noticed that the mercury thread in the thermometer at the higher temperature rose higher still, while the mercury thread in the other thermometer dropped lower still, we would conclude that the film was being run backward.

Making use of a hot body and a cold body, we could arrange to have the heat flow do work. Heat from the hot body could evaporate a liquid and the expanding vapor could push a piston. The vapor could then deliver its heat to the cold body, become liquid again, and the process could continue over and over.

As the work is done and heat flows, the hot body transfers its heat to the evaporating liquid and the vapor, as it condenses, transfers its heat to the cold body. The hot body therefore grows cooler and the cold body gets warmer. As the temperatures approach each other, the rate of heat flow decreases and so does the amount of work done. When the two bodies are at the same temperature, there is then no heat flow and no work is done at all. The bodies are still there, all the heat energy is still there, but there is no longer an uneven distribution of the heat, and therefore no longer any change, any motion, any work.

Once again, the spontaneous change is from uneven distribution of energy to even distribution, from the capacity for change, motion and work, to the absence of such capacity. Again, once such capacity disappears, it does not reappear.

The Second Law of Thermodynamics

Studies on energy usually involve a careful consideration of heat flow and of temperature change because this is the easiest aspect of the subject to handle in the laboratory—and because it was also particularly important at a time when steam engines were the major method of turning energy into work. For this reason the science of energy-change, energy-flow, and the conversion of energy into work was termed "thermodynamics" from Greek words meaning "heat-movement."

The law of conservation of energy is sometimes called "the first law of thermodynamics" because it is the most basic rule governing what will happen and what won't happen in connection with energy.

As for the spontaneous change from an uneven distribution of energy to an even one, that is called "the second law of thermo-dynamics."

The second law of thermodynamics was foreshadowed as early as 1824, when the French physicist Nicolas L. S. Carnot (1796–1832) was the first to study, in careful detail, the heat-flow in steam-engines.

It was not until 1850, however, that the German physicist Rudolf J. E. Clausius (1822–88) suggested that this evening-out process applied to all forms of energy and to all events in the universe. Clausius is therefore usually considered as the discoverer of the second law of thermodynamics.

Clausius showed that a quantity based on the ratio of total heat to temperature in any particular body was important in connection with the evening-out process. He gave the name "entropy" to this quantity. The lower the entropy, the more uneven the energy distribution. The higher the entropy, the more even the energy distribution. Since the spontaneous tendency seems to be invariably for change from an uneven distribution of energy toward an even one, we can say that the spontaneous tendency seems to be for everything to move from a low entropy to a high entropy.

We can put it this way:

The first law of thermodynamics states: the energy content of the universe is constant.

The second law of thermodynamics states: the entropy content of the universe is steadily increasing.

If the first law of thermodynamics seems to imply that the universe is immortal, the second law shows that that immortality is, in a way, worthless. The energy will always be there, but it won't always be able to bring about change, motion, and work.

Someday, the entropy of the universe will reach a maximum and all the energy will be evened out. Then, although all the energy will still be there, no further change will be possible, no motion, no work, no life, no intelligence. The universe will exist but only as the frozen statue of a universe. The film will have stopped rolling and we will be looking forever at a "still."

Since heat is the least organized form of energy and that which lends itself most easily to being evenly spread out, any change from any form of nonheat energy into heat represents an increase in entropy. The spontaneous change is always from electricity to heat, from chemical energy to heat, from radiant energy to heat, and so on.

At maximum entropy, therefore, all forms of energy that can be converted to heat will be, and all parts of the universe will be at the same temperature. This is sometimes called the "heat-death of the universe" and from what I have said so far, it would seem to represent an inevitable and inexorable end.

The ends of the mythic and the scientific universes are thus far different. The mythic universe ends in a vast conflagration and falling apart; it ends in a bang. The scientific universe, if subjected to the heat-death, ends in a long-drawn-out whimper.

The end of the mythic universe always seems to be expected in the near future. The end of the scientific universe by the heat-death route is far off indeed. It is at least a thousand billion years off, perhaps many thousand billions of years off. Considering that the universe is at present only fifteen billion years old according to current estimates, we are clearly only in the infancy of its life.

Yet, although the end of the mythic universe is usually described as violent and near, it is accepted because it brings the promise of regeneration. The end of the scientific universe by heat-death, though it be peaceful and exceedingly far off, seems to include no promise of regeneration but to be final; and apparently that is a hard thing to accept. People search for ways out.

After all, processes that are spontaneous can, nevertheless, be reversed. Water can be pumped upward against its tendency to seek its level. Objects can be cooled below room temperature and kept there in a refrigerator; or heated above room temperature and kept there in an oven. Looked at in that way, it would seem as though the inexorable entropy-increase could be defeated.

Sometimes the process of entropy-increase is described by imagining the universe to be a huge and indescribably intricate clock which is slowly running down. Well, human beings own clocks that can and do run down, but we can always wind them up again. Might there not be some analogous process for the universe?

Indeed, we don't have to imagine entropy-decrease coming about only through the deliberate actions of human beings. Life itself, quite apart from human intelligence, seems to defy the second law of thermodynamics. Individuals die, but new individuals are born and youth is as prevalent now as it always was. Vegetation dies in the winter, but it grows again in the spring. Life has continued on Earth for over three billion years and more and shows no sign of running down. In fact, it shows every sign of winding up, for in all the history of life on Earth, life has been growing more complex both in the case of individual organisms and in the ecological web that binds them all together. The history of biological evolution represents a *vast* decrease in entropy.

Because of this, some people have actually tried to define life as an entropy-decreasing device. If this were true than the universe would never experience a heat-death since wherever life exerts an influence it will automatically act to decrease entropy. As it happens, though, this is all wrong. Life is not an entropy-decreasing device and it cannot by itself avert the heat-death. The thought that it is, and that it can, arises out of wishful thinking and imperfect understanding.

The laws of thermodynamics apply to closed systems. If a pump is used to decrease entropy by moving water uphill, the pump has to be counted in as part of the system. If a refrigerator is used to decrease entropy by cooling objects below room temperature, the refrigerator has to be counted in as part of the system. Nor can the pump or refrigerator be counted in merely as themselves. Whatever they are connected to, whatever their source of power, that, too, must be counted in as part of the system.

Any time human beings and human tools are used to decrease entropy and reverse a spontaneous reaction, it turns out that the human beings and the human tools engaged in the process are suffering an increase in entropy. What's more, the entropy-increase of the human beings and their tools is greater, *invariably,* than the entropy-decrease of that part of the system in which a spontaneous reaction is being reversed. The entropy of the entire system, therefore, increases; *always* increases.

To be sure, a given human being can reverse many, many spontaneous reactions in his life, and many human beings working together have built the enormous technological network that covers the Earth from the pyramids of Egypt and the Great Wall of China right down to the latest skyscraper and dam. Can human beings experience so enormous a rise in entropy and keep right on going?

Again, one can't consider human beings by themselves. They do not form closed systems. A human being eats, drinks, breathes, eliminates wastes, and these are all connections with the outside universe, conduits whereby energy enters or leaves. If you want to consider a human being as a closed system, you have to consider what he eats, drinks, breathes, and eliminates as well.

The entropy of a human being is raised as he reverses spontaneous actions and continually winds up that portion of the unwinding universe he can reach, and, as I said, his entropy-increase more than makes up for the entropy-decrease he brings about. However, a human being continually lowers his entropy again by eating, drinking, breathing, and eliminations. (The lowering is not perfect, of course; eventually each human being dies, no matter how successfully he avoids accident and disease, because of slow entropic rises here and there that cannot be restored.)

However, the increase in entropy in the food, water, air, and elimination portions of the system is, once more, well above the entropy-decrease in the human being himself. For the entire system, there is an entropy increase.

In fact, not only the human being, but all animal life flourishes and maintains its entropy at a low level at the expense of a vast increase in the entropy of its food which, in the last analysis, consists of the vegetation of the Earth. How, then, does the plant world continue to exist? After all, it can't exist for long if its entropy rises continually.

The plant world produces the food and oxygen (the key component of air) that the animal world lives on by a process known as "photosynthesis." It has been doing this for billions for years; but then plant and animal life taken as a whole are not a closed system either. The plants derive the energy that drives their production of food and oxygen from sunlight.

It is therefore sunlight that makes life possible and the sun itself must be included as part of the life-system before the laws of thermodynamics can be applied to life. As it happens, the sun's entropy rises steadily by an amount that far outstrips any entropy-decrease that can be brought about by life. The net change in entropy of the system that includes life *and* the sun is therefore a pronounced and continuing rise. The vast entropy-decrease represented by biological evolution, then, is only a ripple on the tidal wave of entropy-increase represented by the sun, and to concentrate on the ripple to the exclusion of the tidal wave is to completely misinterpret the facts of thermodynamics.

Human beings make use of sources of energy other than the food and oxygen they eat and breathe. They make use of the energy of wind and running water, but both are products of the sun since winds are the product of the uneven heating of the Earth by the sun and running water begins with the sun's evaporation of the ocean.

Human beings make use of burning fuel for energy. But here the fuel may be wood or other plant products, which are based on light from the sun. It may be fat or other animal products, and animals feed on plants. It may be coal, which is the product of past ages of plant growth. It may be petroleum, which is the product of past ages of microscopic animal growth. All these fuels trace back to the sun.

There is energy on Earth that doesn't come from the sun. There is energy in Earth's internal heat and that produces hot springs, geysers, earthquakes, volcanoes, and the shifting of Earth's crust. There is energy in Earth's rotation, which is evidenced in the tides. There is energy in inorganic chemical reactions and in radioactivity.

All these sources of energy produce changes, but in every case the entropy is rising. Radioactive materials are slowly decaying away, and once their heat is no longer added to Earth's internal supply, the earth will cool off. Tidal friction is gradually slowing the Earth's rotation, and so on. Even the sun will eventually run out of its supply of work-producing energy as its entropy rises. And the biological evolution of the last three billion years and more, which seems so remarkable an entropy-decreasing process, has done it on the basis of the rising entropy of all these energy sources and, it would seem, can do nothing to stanch *that* rise.

In the long run, it would appear, nothing can withstand the rising level of entropy or keep it from reaching maximum, at which time the heat-death of the universe will arrive. And if human beings could escape all other catastrophes and somehow still exist trillions of years from now, then will they not finally bow to the inevitable and die with the heat-death?

From all I have said so far, it would seem so.

Movement at Random

Yet there is something disturbing in this picture of the steadily rising entropy-content of the universe, and it shows up as we look backward in time.

Since the entropy-content of the universe is rising steadily, the entropy of the universe must have been less a billion years ago than it is now, and still less two billion years ago, and so on. At some moment, if we go back far enough, the entropy of the universe must have been zero.

Astronomers currently believe that the universe began about 15 billion years ago. By the first law of thermodynamics the energy of the universe is eternal, so when we say that the universe began 15 billion years ago, we don't mean that the energy (including matter) of the universe was then created. It always existed. All we can say is that it was 15 billion years ago that the entropy-clock started ticking and running down.

But what wound it up in the first place?

To answer that question, let's go back to my two examples of spontaneous entropy-increase—the water flowing from a full container to a nearly empty one, and the heat flowing from a hot body to a cold body. I implied that the two are strictly analogous; that heat is a fluid as water is and behaves in the same way. Yet there are problems in that analogy. It is easy, after all, to see why the water in the two containers acts as it does. There is gravity pulling at it. The water, responding to the uneven gravitational pull on itself in the two containers, flows from the full container into the nearly empty one. When the containers each have water reaching the same level, the gravitational pull is equal on both and there is no further motion. But what is it that, analogously to gravity, pulls at heat and drags it from a hot body to a cold body? Before we can answer that, we must ask: What is heat?

In the eighteenth century, heat was actually thought to be a fluid, like water but much more ethereal, and therefore capable of pouring into and out of the interstices of apparently solid objects, much as water can pour into and out of a sponge.

In 1798, however, the American-born British physicist Benjamin Thompson, Count Rumford (1753–1814), studied the production of heat from friction when cannon were being bored, and suggested that heat was actually the motion of very small particles of matter. In 1803, the English chemist John Dalton (1766–1844) worked out the atomic theory of matter. All matter was made up of atoms, he said. From Rumford's point of view it might be the motion of these atoms that represented heat.

About 1860, the Scottish mathematician James Clerk Maxwell (1831–79) worked out the "kinetic theory of gases," showing how to interpret their behavior in terms of the atoms or molecules* that

---

* A molecule is a group of atoms, holding more or less firmly together, and moving as a unit.

made them up. He showed that these tiny particles, moving in any direction at random, and colliding with each other and with the walls of any container housing them, again at random, could account for the rules governing gas behavior that had been worked out over the previous two centuries.

In any sample of gas, the constituent atoms or molecules move in any of a wide range of velocities. The average velocity, however, is higher in hot than in cold gases. In fact, what we call temperature is equivalent to the average velocity of the constituent particles of a gas. (This holds, by extension, for liquids and solids, too, except that in liquids and solids, the constituent particles are vibrating rather than moving bodily.)

For the sake of simplifying the argument that follows, let's suppose that in any sample of matter at a given temperature, all the particles making it up are moving (or vibrating) at the average velocity characteristic of that temperature.

Imagine a hot body (gas, liquid, or solid) brought into contact with a cold body. The particles at the edge of the hot body will collide with those at the edge of the cold body. A fast particle from the hot body will collide with a slow one from the cold body, and the two will then rebound. The total momentum of the two particles stays the same but there can be a transfer of momentum from one body to the other. In other words, the two particles can leave each other with different speeds than those with which they approached.

It is possible that the fast particle will give up some of its momentum to the slow particle, so that the fast particle will, after rebounding, move more slowly while the slow one will, after rebounding, move more quickly. It is also possible that the slow particle will give up some of its momentum to the fast one so that the slow particle will rebound more slowly still, and the fast particle will rebound still faster.

It is just chance that determines in which direction the transfer of momentum will take place, but the odds are that the momentum will transfer from the fast particle to the slow one, that the fast particle will rebound more slowly, and that the slow particle will rebound more quickly.

Why? Because the number of ways in which momentum can transfer from the fast particle to the slow one is greater than the number of ways in which momentum can transfer from the slow particle to the fast one. If all the different ways are equally likely, then there is a greater chance that one of the many possible transfers from fast to slow will be taken rather than one of the few possible transfers from slow to fast.

To see why this is so, imagine fifty poker chips in a jar all identical, labeled with numbers from 1 to 50. Pick one at random and imagine you have picked number 49. That's a high number and represents a fast-moving particle. Put chip 49 back in the jar (that represents a collision) and select another numbered chip at random (that represents the speed at rebound). You might pick 49 again and rebound at the same speed with which you had collided. Or you might pick 50 and rebound even more quickly than you had collided. *Or* you might pick any number from 1 to 48, forty-eight different choices, and in each case rebound more slowly than you had collided.

Having picked 49 to begin with, your chance for rebound at a higher velocity is only 1 out of 50. The chance of rebounding more slowly is 48 out of 50.

The situation would be reversed if you had picked chip number 2 to start with. That would represent a very slow speed. If you then threw it back and picked again, you would have only 1 chance out of 50 to pick a 1 and rebound even more slowly than you had collided, while you would have 48 chances out of 50 to pick any number from 3 to 50 and rebound more quickly than you had collided.

If you imagined ten people each picking poker chip 49 out of a separate jar, and each throwing it back to try their luck again, the chance that every one of them would pick 50 and that every one of them would rebound more quickly than he or she had collided would be 1 in about a hundred million billion. The chances are 2 out of 3, on the other hand, that every single one of the ten would come out with a rebound at a slower speed.

The same thing would happen in reverse if we imagined ten people each picking number 2 and then trying again.

Nor do all these people have to pick the same number. Let us say that a large number of people pick chips and get all sorts of different numbers, but that the average is quite high. If they try again, the average is much more likely to be lower than to be still higher. The more people there are, the more certain it is that the average will be lower.

The same is true of many people picking chips and finding themselves with a quite low average value. The second chance is very likely to raise the average. The more people there are, the more likely the average is to be raised.

In any body large enough to be experimented with in the laboratory, the number of atoms or molecules involved in each is not ten or fifty or even a million, but billions of trillions. If these billions of trillions of particles in a hot body have a high average speed, and if billions of trillions of particles in a cold body have a low average speed, then the odds are tremendous that random collisions among the lot of them are going to bring down the average of the particle velocities in the hot body and bring up the average in the cold body.

Once the average particle-speed is the same in both bodies, then momentum is just as likely to transfer in one direction as in the other. Individual particles may go now faster now slower, but the average speed (and therefore the temperature) will remain the same.

This gives us our answer as to why heat flows from a hot body to a cold body and why both come to the same average temperature and remain there. It is simply a matter of the laws of probability, the natural working out of blind chance.

In fact, that is why entropy continually rises in the universe. There are so many, many more ways of undergoing changes that even out energy distribution than there are those that make it more uneven, that the odds are incredibly high that the changes will move in the direction of increasing entropy through nothing more than blind chance.

The second law of thermodynamics, in other words, does not describe what *must* happen, but only what is *overwhelmingly likely* to happen. There's an important difference there. If entropy *must* increase then it can *never* decrease. If entropy is merely overwhelmingly likely to increase, then it is overwhelmingly unlikely to decrease, but eventually, if we wait long enough, even the overwhelmingly unlikely may come to pass. In fact, if we wait long enough, it *must* come to pass.

Imagine the universe in a state of heat-death. We might think of it as a vast three-dimensional sea of particles, perhaps without limit, engaged in a perpetual game of collision and rebound, with individual particles moving more quickly or more slowly, but the average remaining the same.

Every once in a while, a small patch of neighboring particles develops a rather high average speed among themselves, while another patch, some way off, develops a rather low average speed. The overall average in the universe doesn't change, but we now have a patch of low entropy and a small amount of work becomes possible until the patch evens out, which it will do after a while.

Every once in a longer while, there is a larger unevenness produced by these random collisions, and again in an even longer while, a still larger unevenness. We might imagine that every once in a trillion trillion trillion years so large an unevenness is produced that there is a patch the size of a universe with a very low entropy. It takes time for a universe-sized patch of low entropy to even out again; a very long time—a trillion years or more.

Perhaps that is what happened to us. In the endless sea of heat-death, a low-entropy universe found itself suddenly in existence through the workings of blind chance, and in the process of raising its entropy and evening itself out again, it differentiated into galaxies and stars and planets, brought forth life and intelligence, and here we are, wondering about it all.

Thus, the ultimate catastrophe of heat-death may be followed by regeneration after all, just as the violent catastrophes described in Revelation and Ragnarok were.

Since the first law of thermodynamics would seem to be absolute, and the second law of thermodynamics would seem to be only statistical, there is the chance of an infinite succession of universes, separated each from each by unimaginable eons of time, except that there will be no one and nothing to measure the time and no way, in the absence of rising entropy, to measure it even if instruments and inquiring minds existed. We might, therefore, say that the infinite succession of universes was separated by timeless intervals.

And how does that affect the tale of human history?

Suppose human beings have somehow survived all other possible catastrophes and that our species is still alive trillions of years from now when the heat-death is upon the universe. The rate of entropy increase drops steadily as the heat-death approaches and patches of comparatively low entropy (patches that are small in volume compared to the universe, but very large on the human scale) would linger here and there.

If we assume that human technology has advanced more or less steadily over a trillion years, human beings should be able to take advantage of these patches of low entropy, discovering them and exploiting them as we now discover and exploit gold mines. These patches could continue running down, and supporting humanity in the process, for billions of years. Indeed, human beings might well discover new patches of low entropy as they form by chance in the sea of heat-death, and exploit those, too, in this way continuing to exist indefinitely, although under constricted conditions. Then, finally, chance will provide a patch of low entropy of universe size and human beings will be able to renew a relatively boundless expansion.

To take the absolute extreme, human beings may do as I once described them as doing in my science-fiction story "The Last Question," first published in 1956, and seek to discover methods for bringing about a massive decrease in entropy, thus averting the heat-death, or deliberately renewing the universe if the heat-death is already upon us. In this way, humanity might become essentially immortal.

The question is, however, whether human beings will still be in existence at a time when the heat-death becomes a problem, or whether some earlier catastrophe of another kind is sure to wipe us out.

That is the question to which the rest of the book will seek an answer.

Chapter 3

The Closing
of the Universe

The Galaxies

So far, we have been discussing the manner in which it would seem the universe ought to behave in accordance with the laws of thermodynamics. It is time we took a look at the universe itself in order to see whether that would cause us to modify our conclusions. To do this, let us step back and try to look at the contents of the universe as a whole, generally; something we have only been able to do in the twentieth century.

Throughout earlier history, our views have been restricted to what we could see of the universe, which turned out to be very little. At first the universe was merely a small patch of Earth's surface over which the sky and its contents were merely a canopy.

It was the Greeks who first recognized the Earth to be a sphere and who even gained a notion of its true size. They recognized that the sun, moon, and the planets moved across the sky independently of the other objects, and supplied each of them with a transparent sphere. The stars were all crowded into a single outermost sphere and were considered merely background. Even after Copernicus sent the Earth hurtling around the sun, and the coming of the telescope revealed interesting details concerning the planets, the consciousness of human beings did not really extend beyond the solar system. As late as the eighteenth century, the stars were still little more than background. It was only in 1838 that the German astronomer Friedrich Wilhelm Bessel (1784–1846) determined the distance of a star and the scale of interstellar distances was established.

Light travels at the speed of nearly 300,000 kilometers (186,000 miles) per second and in one year light will therefore travel 9.44 trillion kilometers (5.88 trillion miles). That distance is a light-year, and even the nearest star is 4.4 light-years away. The average distance between stars in our neighborhood of the universe is 7.6 light-years.

The stars do not seem to be spread out through the universe in all directions alike. In a circular band around the sky there are so many stars that they fade off into a dimly luminous fog called the "Milky Way." In other areas of the sky there are, by comparison, few stars.

It became clear in the nineteenth century, therefore, that the stars were arranged in the shape of a lens, much wider than it is thick, and thicker in the middle than toward the rim. We now know that the lens-shaped conglomeration of stars is 100,000 light-years across in its widest dimension and that it contains perhaps as many as 300 billion stars, with an average mass of perhaps half that of our sun. This conglomeration is called the "Galaxy," from the Greek expression for "Milky Way."

Throughout the nineteenth century, it was assumed that the Galaxy was just about all there was to the universe. There didn't seem to be anything in the sky that was distinctly outside it except for the Magellanic clouds. These were objects in the southern sky (invisible from the North Temperate Zone) which looked like detached fragments of the Milky Way. They turned out to be small conglomerations of stars, only a few billion in each, that lay just outside the Galaxy. They could be considered small satellite-galaxies of the Galaxy.

Another suspicious object was the Andromeda nebula just visible as a dim and fuzzy object to the naked eye. Some astronomers thought it was just a bright cloud of gas that was part of our own galaxy, but if so, why were there no stars visible inside it to serve as the source of the light? (Stars were visible in the case of other bright clouds of gas in the Galaxy.) Then, too, the nature of its light seemed that of starlight and not that of luminous gas. Finally, novas (suddenly brightening stars) appeared in it with surprising frequency, novas that would not be visible at their ordinary brightness.

There was good reason to argue that the Andromeda nebula was a conglomeration of stars, as large as the Galaxy, that was so far distant that none of the individual stars could be made out—except that occasionally, one of its stars, brightening for some reason, would become bright enough to see. The most vigorous champion of this view was the American astronomer Heber Doust Curtis (1872–1942), who made a special study of the novas in the Andromeda nebula in 1917 and 1918.

Meanwhile, in 1917, a new telescope with a 100-inch mirror (the largest and best the world had seen up to that time) was installed on Mount Wilson, near Pasadena, California. Using that telescope, the American astronomer Edwin Powell Hubble (1889–1953) finally managed to make out individual stars on the outskirts of the Andromeda nebula. It was definitely a conglomeration of stars of the size of our galaxy and since then it has been called the Andromeda galaxy.

We now know that the Andromeda galaxy is 2.3 million light-years away from us, and that there are vast numbers of other galaxies stretching out in every direction for ten billion light-years and more. Therefore, if we consider the universe as a whole, we would have to consider it as a large conglomeration of galaxies, fairly evenly distributed through space, with each galaxy containing anywhere from a few billion to a few trillion stars.

The stars within a galaxy are held together by their mutual gravitational pull and each galaxy turns as the various stars move in orbits about the galactic center. Thanks to gravity, galaxies can remain intact and can retain their identities over many billions of years.

What's more, it is common for neighboring galaxies to form groups or clusters in which all are bound to each other by their mutual gravitational pull. For instance, our own galaxy, the Andromeda galaxy, the two Magellanic clouds, and over twenty other galaxies (most of them quite small) make up the "local group." Among the other galactic clusters we can see in the sky, some are much more enormous. There is one cluster in the constellation Coma Berenices, about 120 million light-years distant, that is made up of about 10,000 individual galaxies.

It may be that the universe is made up of about a billion galactic clusters, each with an average of about a hundred members.

The Expanding Universe

Even though the galaxies are enormously distant, some interesting things can be learned about them from the light that reaches us from them.

The visible light that reaches us from any hot object, be it a vast cluster of galaxies, or a bonfire, is made up of a variety of wavelengths from the shortest that will affect the retina of our eye to the longest. There are instruments that can sort out these wavelengths into bands stretching, in order, from the shortest to the longest. Such bands are called "spectra" (singular, "spectrum").

The wavelengths affect our eyes in such a way as to be interpreted as colors. Visible light of the shortest wavelength appears to us as violet. As the wavelengths grow longer, we see, in order, blue, green, yellow, orange, and red. This is the familiar rainbow, and, indeed, the rainbow we see in the sky after a shower is a natural spectrum.

When the light of the sun or of other stars is spread out into a spectrum, some of the wavelengths of light are missing. These have been absorbed en route by the relatively cool gases in the upper atmosphere of the sun (or of the other stars). These missing wavelengths show up as dark lines crossing the various colored bands of the spectrum.

Each type of atom in the atmosphere of a star absorbs wavelengths characteristic for itself and for no other. The location of the characteristic wavelengths in the spectrum can be determined accurately in the laboratory for each type of atom and, from the dark lines in the spectrum of any star, information on the chemical composition of that star can be obtained.

As long ago as 1842, the Austrian physicist Christian Johann Doppler (1803–53) showed that when a body emitted sound of a certain wavelength, that wavelength increased if the body were moving away from us as it emitted the sound, and decreased if the body were moving toward us. In 1848, the French physicist Armand H. L. Fizeau (1819–96) applied this principle to light.

By this Doppler-Fizeau effect, all the wavelengths of light emitted by a star that is moving away from us are longer than they would be if they were emitted by a stationary object. This includes the dark lines particularly which are shifted toward the red end of the spectrum (the "red shift") as compared to where they would ordinarily be. In the case of a star moving toward us, the wavelengths, including the dark lines, shift toward the violet end of the spectrum.

By determining the position of particular dark lines in the spectrum of a specific star, not only can it be decided whether that star is receding from us or approaching us, but at what rate—since the faster the star is either receding or approaching, the greater the shift in the dark lines. This shift was used for the first time in 1868, when the English astronomer William Huggins (1824–1910) detected a red shift in the spectrum of the star, Sirius, and determined that it was receding from us at a moderate speed. As more and more stars were tested in this way, it turned out, not surprisingly, that some were approaching and some receding from us, as was to be expected if the Galaxy was, as a whole, neither moving toward us or away from us.

In 1912, the American astronomer Vesto Melvin Slipher (1875–1969) began a project of determining the dark-line shift of the various galaxies (even before the little bits of cloudy light had been definitely recognized as galaxies).

It might be supposed that the galaxies, too, would show some recessions and some approaches as the stars do; and, indeed, this is true for the galaxies of our local group. For instance, the first galaxy studied by Slipher was the Andromeda galaxy and it turned out to be approaching our own galaxy at a speed of about 50 kilometers (32 miles) per second.

The galaxies outside our local group, however, showed a puzzling uniformity, Slipher and those who followed him found that, in every case, the light from the galaxies showed a red-shift. One and all were receding from us, and at unusually high speeds. Whereas the stars of our galaxy moved about relative to each other at speeds of some tens of kilometers per second, even the nearer galaxies outside our local group were receding from us at speeds of some hundreds of kilometers per second. Furthermore, the fainter a galaxy (and, presumably, the more distant it was) the more rapidly it receded from us.

By 1929, Hubble (who five years before had detected stars in the Andromeda galaxy and settled its nature) was able to show that the speed of recession was proportional to distance. If galaxy A was three times as far from us as galaxy B was, then galaxy A receded from us at three times the speed that galaxy B did. Once this was accepted, the distance of a galaxy could be determined merely by measuring its red shift.

But why should all the galaxies be receding from us?

To explain this universal recession without assuming some special quality for ourselves, it was only necessary to accept as fact that the universe was expanding, and that the distance between all neighboring galactic clusters was constantly increasing. If that were so, then from a viewing station within *any* galactic cluster, and not just from one within our own, all the other galactic clusters would seem to be receding at a rate that increased steadily with distance.

But why should the universe be expanding?

If we imagined time to be moving backward (that is, if we imagine ourselves to have taken a movie film of the expanding universe and then to be running the film backward) the galactic clusters would seem to be approaching each other and, eventually, they would coalesce.

The Belgian astronomer Georges Lemaître (1894–1966) suggested in 1927 that at some point in time long ago all the matter of the universe was squeezed into a single object which he called the "cosmic egg." It exploded and the galaxies were formed out of the fragments of the explosion. The expanding universe is expanding because of the force of that ancient explosion.

The Russian-American physicist George Gamow (1904–68) called this primordial explosion the "big bang" and that is the phrase everyone uses now. It is the big bang that astronomers now think took place some 15 billion years ago. The entropy of the cosmic egg was very low and from the moment of the big bang that entropy has been increasing and the universe has been running down, as described in the previous chapter.

Did the big bang really take place?

The farther we penetrate into the vast distances of the universe, the farther back into time we are peering. After all, it takes light time to travel. If we could see something that was a billion light-years away, then the light we see would have taken a billion years to reach us and the object we see would be as it was a billion years ago. If we could see something that was 15 billion light-years away, then we would see it as it was 15 billion years ago at the time of the big bang.

In 1965, A. A. Penzias and R. W. Wilson of Bell Telephone Laboratories were able to show that there was a faint glow of radio waves coming evenly from every part of the sky. This radio-wave background seems to be the radiation of the big bang reaching us from across fifteen billion light-years of space. This finding has been accepted as strong evidence in favor of the big bang.

Will the universe be expanding forever as a result of that enormous primordial explosion? I will discuss this possibility shortly, but for now let us suppose that the universe will indeed expand forever. In that case, how will it affect us? Does the indefinite expansion of the universe constitute a catastrophe?

Visually, at least, it does not. Without exception, everything we see in the sky with the naked eyes, including the Magellanic clouds and the Andromeda galaxy, are part of the local group. All parts of the local group are held together gravitationally and are not taking part in the general expansion.

What it amounts to, then, is that though the universe may expand forever, our view of the heavens without a telescope will not change because of that. There will be other changes for other reasons, but our local group, containing over half a trillion stars altogether, will stay put.

As the universe expands, astronomers will have greater and greater difficulty in making out the galaxies outside the local group, and will eventually lose them altogether. All the galactic clusters will recede to such a distance that they will be moving away from us at such speeds as to be unable to affect us in any way. Our universe will then consist of the local group only and will be only one fifty-billionth as large as it is now.

Would this vast shrinkage in the size of our universe constitute a catastrophe? Not directly, perhaps, but it would affect our ability to deal with the heat-death.

A smaller universe would have less of a chance to form a large area of low entropy and it could never, by random processes, form the kind of cosmic egg that started our universe. There wouldn't be enough mass for that. To make an analogy, there would be far less chance of finding a gold mine if we were only to dig in our own backyard, than if we were allowed to dig anywhere on Earth's surface.

Thus, the indefinite expansion of the universe vastly decreases the possibility that the human species might survive the heat-death—if it lasts that long to begin with. In fact, one might be strongly tempted to predict that it won't; the combination of infinite expansion and heat-death would be too much for the human species to defeat even at the most optimistic interpretation of events.

Yes, this is not all, either. Is it possible that the recession of the galactic clusters alters the properties of the universe in such a way as to bring about a catastrophe more immediate than failure to survive the heat-death?

Some physicists speculate that gravitation is the product of all the mass in the universe working in cooperation and is not merely the product of individual bodies. The more the total mass of the universe is concentrated into a smaller and smaller volume, the more intense the gravitational field produced by any given body. Likewise the more the mass is dissipated into larger and larger volumes, the weaker the gravitational force produced by any given body.

Since the universe is expanding, the mass of the universe is being spread out over a larger and larger volume and the intensity of the individual gravitational fields produced by the various bodies of the universe should, by that line of thought, be decreasing slowly. This possibility was first suggested in 1937 by the English physicist Paul A. M. Dirac (1902–).

It would be a very slow decrease and its effects would not be noticeable to ordinary individuals for many millions of years, but gradually the effects would pile up. The sun, for instance, is held together by its powerful gravitational field. As the gravitational force grew weaker, the sun would slowly expand and grow cooler, and so would all the other stars. The sun's grip on Earth would weaken, so that Earth's orbit would spiral very slowly outward. Earth itself, with its own gravity weakening, would expand slowly, and so on. We might then face a future in which Earth's temperature, thanks to a cooling and more distant sun, might drop and freeze us out. This and other effects might bring us to an end before we ever reached the heat-death.

So far, however, scientists have failed to find any clear-cut sign that gravitation is weakening with time, or that, in the course of Earth's past history, it was ever significantly stronger.

It is perhaps too soon to talk and we should wait for further evidence before allowing ourselves to be too certain in the matter one way or the other, but I can't help but feel that the notion of a weakening gravitational force is untenable. If it were so and Earth would be growing cooler in the future, then by the same token it would have been hotter in the past, and there is no sign of that. Then, too, the gravitational fields generally would be stronger and stronger as we move into the past and, at the time of the cosmic egg, they would be so strong that it seems to me the cosmic egg could never have exploded and hurled fragments outward against the pull of that unimaginably intense gravitational field.*

Until evidence to the contrary is found, then, it seems sensible to suppose that the indefinite expansion of the universe will not affect the properties of our own portion of the universe. The expansion is not likely, therefore, to bring about a catastrophe earlier than the time when it would be very unlikely that humanity could survive the heat-death.

The Contracting Universe

But wait. How sure can we be that the universe will expand forever just because it is expanding now?

Suppose, for instance, that we are watching a thrown ball moving upward from the surface of the Earth. It is moving upward steadily, but at a speed that is steadily slowing. We know that eventually its upward speed will be reduced to zero and that it will then begin to move downward, faster and faster.

The reason for this is that the Earth's gravitational pull drags downward on the ball inexorably, first diminishing its initial upward impulse until it is gone, and then adding continually to its eventual downward plunge. If the ball were thrown upward more rapidly to begin with, it would take longer for the gravitational pull to counteract that initial impulse. The ball would manage to reach a greater height before it came to a halt and began to fall again.

We might imagine, then, that no matter how rapidly we hurled the ball upward to begin with, it would eventually halt and return under gravity's inexorable drag. In fact, we have the folk-saying, "What goes up must come down." This would be true if the gravitational pull were constant all the way up, but it isn't.

The pull of Earth's gravity decreases as the square of the distance from the Earth's center. An object on the Earth's surface is roughly 6,400 kilometers (4,000 miles) from its center. An object 6,400 kilometers above its surface would be twice as far from its center and would be under a gravitational pull only ¼ that on the surface.

An object can be thrown upward with so great a velocity that as it moves upward the gravitational pull decreases so rapidly that it is

---

* Indeed, as we shall soon see, there is some question as to whether the big bang could take place even with our presently intense gravitational fields.

never strong enough to slow that speed down to zero. Under those circumstances, the object does not come down but leaves the Earth forever. The minimum speed at which that happens is the "escape velocity," and for Earth the escape velocity is 11.23 kilometers (6.98 miles) per second.

The universe may be considered as having an escape velocity, too. The galactic clusters attract each other gravitationally, but as a result of the force of the big bang explosion, are moving apart against the pull of gravity. This means that we can count on the gravitational pull to slow the expansion, little by little, and possibly bring it to a halt. Once this happens the galactic clusters will, under the force of their own gravitational attractions, begin to approach each other and there will thus come into being a contracting universe. As the galactic clusters move away from each other, however, the gravitational pull of each on its neighbors decreases. If the expansion is fast enough, the pull decreases at such a rate that it can never succeed in bringing the expansion to a halt. The minimum rate of expansion required to prevent that halt is the escape velocity for the universe.

If the galactic clusters are separating from each other at more than the escape velocity, they will continue to separate forever and the universe will expand forever until it reaches the heat-death. It will be an "open universe" of the kind we were discussing earlier in the chapter. If the galactic clusters are separating at less than the escape velocity, however, the expansion will gradually come to a halt. A contraction will then eventually begin and the universe will reform the cosmic egg which will then explode in a new big bang. It will be a "closed universe" (sometimes called an "oscillating universe").

The question, then, is whether or not the universe is expanding at a rate that is beyond the escape velocity. We know the rate of expansion and if we also knew the value of the escape velocity, we would have the answer.

The escape velocity depends on the gravitational attraction of the galactic clusters on each other, and this depends on the mass of the individual galactic clusters and how far apart they are from each other. Of course, different galactic clusters come in different sizes and some neighboring galactic clusters are farther apart than others.

What we can do, therefore, is to imagine all the matter in all the galactic clusters smeared out evenly over the universe. We could then determine the average density of matter in the universe. The higher the average density of matter, the greater the escape velocity, and the more likely it is that the galactic clusters are not separating from each other quickly enough to escape and that sooner or later the expansion will come to a halt and turn into a contraction.

As nearly as we can tell now, if the average density of the universe were such that a volume equal to a good-sized living room would hold enough matter to make up the equivalent of 400 hydrogen atoms, that would represent a high enough density to keep the universe closed under the present rate of expansion.

As far as we know, however, the actual average density of the universe is only one one-hundredth that quantity. From certain indirect evidence, including the quantity of deuterium (a heavy variety of hydrogen) in the universe, most astronomers are convinced that the average density *can't* be much higher than this. If this is so, the gravitational pull of the galactic clusters on each other is far too small to bring the expansion of the universe to a halt. The universe is therefore open and expansion will continue to the final heat-death.

Except that we're not utterly sure of the average density of the universe. The density is equal to the mass per volume and although we know the volume of a given section of the universe reasonably well, we are not so sure of the mass of that section.

We have ways of calculating the masses of the galaxies themselves, but we are not so good at measuring the mass of the thin scattering of stars, dust, and gas on the far outskirts of galaxies and in between the galaxies. It may be that we are grossly underestimating the mass of this nongalactic material.

Indeed, in 1977, Harvard astronomers studying X-rays from space reported they found indications that some galactic clusters are surrounded by haloes of stars and dust that possess up to five to ten times the mass of the galaxies themselves. Such halos, if common, would add substantially to the mass of the universe and make the possibility of an open universe very uncertain indeed.

One indication that the possibility of a much higher mass for the universe is to be taken seriously rests in the galactic clusters themselves. In many cases, when the mass of the galactic clusters is calculated on the basis of the masses of the component galaxies, it turns out there is not a high enough general gravitational interaction to hold the cluster together. The individual galaxies ought to separate and disperse for they move at rates that are greater than the apparent escape velocity for the cluster. And yet those galactic clusters seem gravitationally bound. The natural conclusion is that astronomers are underestimating the total mass of the clusters; that there is mass outside the galaxies proper which they are not counting in.

In short, while the balance of the evidence is still strongly in favor of an open universe, the chances for it are decreasing somewhat. The chance that there is enough mass in the universe to make it closed and oscillating, while still small, is growing.*

Yet, does a contracting universe make sense? It would bring all the galaxies closer and closer together and, in the end, reform the low-entropy cosmic egg. Does that not mean that a contracting universe defies the second law of thermodynamics? It must contradict it, certainly, but we need not look at this as defiance.

The second law of thermodynamics is, as I have said before, merely a generalization of common experience. We observe that as we study the universe under all sorts of conditions, the second law never seems to be broken; from that we conclude that it cannot be broken.

That conclusion may go too far. After all, no matter how we vary the conditions of experiment and the places we observe, one thing we cannot vary. All the observations we make, from Earth itself to the farthest galaxy we can detect, and all the conditions of experiment we can devise—all, without exception—take place in an expanding universe. Therefore, the most general statement we can make is that the second law of thermodynamics can never be broken in an expanding universe.

On the basis of our observations and experiments we can say exactly nothing about the relationship between entropy and a contracting universe. We are perfectly free to suppose that as the expansion of the universe slows down, the drive to increase entropy becomes less compelling; that as the compression of the universe begins, the drive to decrease entropy begins to become compelling.

We might suppose, then, that in a closed universe, entropy would generally increase during the stage of expansion and, very likely before the heat-death stage is reached, there would be a turnabout and entropy would then decrease during the stage of contraction. The universe, like a carefully-tended watch, then, is rewound before it can run down altogether and, in this fashion, goes on, as nearly as we can tell, forever. But then because the universe goes on, cyclicly,

* If I may again intrude a personal opinion, I feel that an open universe is not really possible for reasons I will explain in the next chapter. I feel that if we are only patient, astronomers will find the missing mass, or whatever other properties are required, and will accept a closed universe.

forever without a heat-death, can we be sure that that means life will continue forever? Might there not be some periods in the cycle during which life is impossible?

For instance, it certainly seems inevitable that the explosion of the cosmic egg is likely to be a condition inimical to life. The entire universe (consisting of only the cosmic egg) is, at the instant of explosion, at a temperature of many trillions of degrees and it is not until some time after the explosion that temperatures have cooled sufficiently for matter to form and to clump together into galaxies, for planetary systems to form, and for life to evolve on suitable planets.

It may not be till about a billion years after the big bang that galaxies, stars, planets, and life could exist in the universe. Assuming that the contraction repeats the history of the universe in reverse, we would expect that for a billion years before the formation of the cosmic egg, life, planets, stars, and galaxies would be impossible.

There is thus a two-billion-year period in each cycle, centered about the cosmic egg, in which life is impossible. In each cycle after this period new life may form, but it will have no connection with the life in the earlier cycle and it will come to an end before the next cosmic egg and have no connection with the life in the later cycle.

Consider: There may not be much less than a trillion stars in the universe. All are pouring energy ceaselessly into the universe generally and have been doing so for 15 billion years. Why has not all this energy served to warm all the cold bodies of the universe—such as planets like our own Earth—to blazing heat and made life impossible?

There are two reasons why this doesn't happen. In the first place, all the galactic clusters are moving apart in an expanding universe. That means the light reaching any of the galactic clusters from all the others undergoes red-shifts to varying degrees. Since the longer the wavelength the lower the energy content of light, the red-shift means a decrease in energy. Therefore, the radiation being emitted by all the galaxies is less energetic than might be thought.

Secondly, the space available within the universe is increasing rapidly as it expands. Space is, in fact, growing more voluminous, faster than the energy being poured into it can fill it. Therefore, far from heating up, the universe has been steadily dropping in temperature since the big bang, and is now at a general temperature of only about 3 degrees above absolute zero.

The situation would, of course, reverse itself in a contracting universe. All the galactic clusters would be moving together and that would mean that the light reaching any one galactic cluster from all the others would undergo violet shifts to varying degrees and would be far more energetic than it is now. Then, too, the space available within the universe would be decreasing rapidly so that the radiation would fill it up rather more rapidly than one might expect. A contracting universe would therefore grow steadily hotter and, as I said, by a billion years before the formation of the cosmic egg would be too hot for anything like life to exist.

How long will it be before the next cosmic egg?

That is impossible to tell. Again, it depends on the total mass of the universe. Suppose the mass is great enough to guarantee a closed universe. The greater the mass beyond the minimum required, the stronger the general gravitational field of the universe and the more quickly the present expansion will be brought to a halt and the whole contracted into another cosmic egg.

Since, however, the present figure for the total mass is so small, it seems likely that if it can be raised high enough to insure a closed universe, it is likely to be raised just *barely* high enough. That means the rate of expansion will slow only very gradually with time, and when it is almost halted the last dregs will disappear only very slowly under the pull of a gravitational field just barely large enough to do the job, and the universe will then begin to contract in a very long-drawn-out fashion.

We are living through a relatively short rapid-expansion period and there will someday be a relatively short rapid-contraction period, each lasting only a few dozen billion years; in between will be a long period of a virtually static universe.

We might suppose as a bare guess that the universe will come to a halt about halfway to the heat-death, say after half a trillion years and that it will then be another half a trillion years before the next cosmic egg. In that case the human species has its choice of waiting a trillion years for the heat-death, if the universe is open, or a trillion years for the next cosmic egg, if the universe is closed.

Both seem ultimate catastrophes, but of the two, the cosmic egg is the more crescendo-like, the more violent, the more Revelation-Ragnarok, and the less easily avoided. The human species might prefer the first, but I suspect that what it will get—always assuming it survives long enough to get either—is the latter.*

* And yet, not to end on an altogether lugubrious note, the science-fiction writer Poul Anderson in his novel *Tau Zero* describes a spaceship and crew witnessing, and surviving, the formation and explosion of a cosmic egg—and does so in remarkably plausible detail.

Chapter 4

The Collapse of Stars

Gravitation

In considering the alternate catastrophes of heat-death and of the cosmic egg, we have been dealing with the universe as a whole and have treated it as though it were some kind of more or less smooth sea of thin matter, all of which was gaining entropy and expanding toward heat-death, or all of which was losing entropy and contracting toward a cosmic egg. We have assumed that all parts of it were suffering the same fate in the same way and at the same time.

The fact is, though, that the universe is not smooth at all unless it is viewed from a huge distance and in a very general way. Viewed at close quarters and in detail it is very lumpy indeed.

To begin with, it contains *at least* ten billion trillion stars and the conditions in or near a star are enormously different from conditions far away from a star. What's more, in some places stars are strewn very thickly, while in others they are distributed thinly, and in still others they are virtually absent. It is quite possible then that events in some parts of the universe are quite different from events in others and that, for instance, while the universe as a whole is expanding, parts of it are contracting. We must consider this possibility since it may be that this difference in behavior can lead on to still another kind of catastrophe.

Let us begin by considering Earth, which has been formed out of roughly six trillion trillion kilograms of rock and metal. The nature of its formation was governed to a large extent by the gravitational field generated by all that mass. Thus, the material of the Earth, as it pulled together through the action of the gravitational field, was forced as close to the center as it could get. Every bit of the Earth moved toward the center until some other bit physically blocked the path. In the end, every bit of the Earth was as close to the center as it could get, so that the whole planet had minimum potential energy.

In a sphere, the distance of the various parts of the body from its center is, on the average, less than it would be for any other geometrical shape; so the Earth is a sphere. (So are the sun and the moon and all other sizable astronomical bodies, barring special conditions.)

What's more, the Earth, shaped by gravitation into a sphere, is tightly packed. The atoms that make it up are in contact. In fact, as one considers the situation deeper and deeper below the Earth's surface, the atoms are more and more compressed by the weight of the layers of material above (this weight representing the pull of gravitation).

Even at the center of the Earth, however, the atoms, though substantially compressed, remain intact. Because they are intact they resist the further action of gravity. Earth collapses no further but remains a sphere 12,750 kilometers (7,900 miles) in diameter and, provided it is left entirely to itself, will remain so indefinitely.

Stars cannot be treated in quite the same way, however, for they have masses anywhere from ten thousand to ten million times that of the Earth and that makes a difference.

Consider the sun, for instance, which has a mass 330,000 times that of the Earth. Its gravitational field is 330,000 times as great, therefore, and when the sun formed, the pull that dragged it into a sphere was that much more powerful. Under that enormous pull, the atoms at the center of the sun, trapped under the colossal weight of the upper layers, broke down and were smashed.

This can happen since the atoms are not at all analogous to tiny billiard balls, as was thought in the nineteenth century. They are, instead, for the most part, fluffy shells of electron waves with very little mass, and, at the center of those shells, there is a tiny nucleus that contains almost all the mass. The nucleus has a diameter only 1/100,000 that of the intact atom. An atom is rather like a ping-pong ball with an invisibly small and very dense metal pellet floating in the center.

Under the pressure of the upper layers of the sun, then, the electron-shells of the atoms at the sun's core smash and the tiny nuclei at the atoms' centers are liberated. The isolated nuclei and the electron-shell fragments are so much smaller than the intact atoms that under its own powerful gravitational pull the sun could shrink to surprisingly small dimensions—but it doesn't.

That this shrinkage does not happen is because of the fact that the sun—and other stars as well—is mostly hydrogen. The hydrogen nucleus at the center of the hydrogen atom is a single subatomic particle called a "proton" which carries a positive electric charge. Once the atoms are smashed, the bare protons can move freely and can approach each other much more closely than they could when each was surrounded by electron shells. Indeed the protons can not only approach each other, but they can collide with great force, since the energy of the gravitational pull is converted to heat as the material of the sun comes together and coalesces so that the center of the sun is at a temperature of about 15 million degrees.

The protons, when they collide, sometimes stick together instead of rebounding, thus initiating a "nuclear reaction." In the process of such nuclear reactions some protons lose electric charge to become "neutrons" and, eventually, a nucleus made up of two protons and two neutrons is formed. This is the nucleus of the helium atom.

The process (the same as that which goes on in an earthly hydrogen bomb, but enormously magnified in power) produces vast quantities of heat which turns the entire sun into a glowing ball of incandescent gas and keeps it that way for a long, long time.

While the Earth is kept from contracting any smaller than it is by the resistance of intact atoms, the sun is kept from it by the expansive effect of the heat developed by the nuclear reactions in its interior. The difference is that whereas the Earth can retain its size indefinitely since the intact atoms, left to themselves, will always remain intact, the sun cannot. The Sun's size depends on the continuous production of heat in the center, which in turn depends on a continuous series of nuclear reactions to produce that heat, which in turn depend on a continuing supply of hydrogen, the fuel for such reactions.

But there is only so much hydrogen. Eventually, given enough time, the hydrogen of the sun (or of any star) will dwindle below some critical amount. The nuclear reaction rate will dwindle and so will the energy. There will be insufficient heat to keep the sun (or any star) distended and it will begin to contract. The contraction of a star has important gravitational consequences.

The gravitational pull between any two objects increases as the distance between their centers deceases; increasing, in fact, as the square of the change in distance. If you are at a certain great distance from the Earth and cut that distance in half, the Earth's pull on you increases by 2 × 2, or 4 times. If you cut that distance to one-sixteenth, the Earth's pull on you increases by 16 × 16, or 256 times.

At the present moment, you are on the surface of the Earth and the strength of Earth's gravitational pull on you depends on its mass, your mass, and the fact that you are 6,378 kilometers (3,963 miles) from Earth's center. You can't very well change the Earth's mass significantly, and you may not want to change your own, but what if you imagine changing your distance from Earth's center.

You can, for instance, move closer to Earth's center by boring (in imagination) through the substance of the Earth itself. Well, you may think, the gravitational pull on you increases as you come closer to Earth's center.

No! The dependence of gravitational pull on distance from the center of the attracting body holds only if you are outside the body. Only then can we treat all the mass of the body as though it were concentrated at the center, in calculating gravitational pulls.

If you burrow into the Earth, only the part of the Earth that is nearer the center than you are will attract you toward the center. The part of the Earth that is farther from the center than you are does not contribute to the gravitational pull. Consequently, as you burrow into the Earth, the gravitational pull on you *decreases*. If you were to reach the very center of the Earth (in imagination) there would be no pull toward the center at all, since there would be nothing closer to the center to pull at you. You would be subjected to zero-gravity.

Suppose, however, that the Earth were to contract to only half its radius while retaining all its mass. If you were far away in a spaceship, that wouldn't affect you. The Earth's mass would still be what it was, as would your mass, and your distance from the Earth's center. Whether the Earth expanded or contracted, its gravitational pull on you would not alter (as long as it didn't expand so far that it engulfed you within its substance—in which case its gravitational pull on you would decrease).

Suppose, though, you were standing on the surface of Earth as it began to contract and you stayed on the surface of the Earth during the contraction process. The Earth's mass, and yours, would stay the same, but your distance from the Earth's center would decrease by a factor of 2. You would still be outside the Earth itself, and all the mass of the Earth would be between you and its center, so that the gravitational pull of Earth upon you would increase by a factor of 2 × 2, or 4. In other words, the *surface gravity* of the Earth would increase as it contracted.

If the Earth continued to contract without losing mass and if you continued to remain on the surface, the gravitational pull on you would increase steadily. If the Earth were imagined to decrease to a point of zero diameter (while retaining its mass) and you were standing on that point, the gravitational pull on you would be infinite.

This is true for any body with mass, however large or however small. If you or I or even a proton were compressed more and more, the gravitational pull upon your surface or my surface or a proton's surface would increase endlessly. And if you or I or a proton were reduced to a point of zero diameter while retaining all the original mass, the surface gravity, in each case, would become infinite.

Black Holes

Of course, the Earth is never likely to contract to a smaller size than it is now, as long as it remains in its present condition. Nor will anything that is smaller than the Earth contract. Even objects somewhat larger than the Earth—Jupiter, for instance, which has 318 times the mass of the Earth—will never contract as long as they are left to themselves.

Stars will, however, contract eventually. They have much more mass then planets do, and their very powerful gravitational field will force contraction upon them once their nuclear fuel falls below the critical point and there is no longer sufficient heat produced to counteract the gravitational pull. How far the contraction proceeds depends upon the intensity of the gravitational field of the contracting body, and therefore upon its mass. If the body is massive enough, there are no limits to the contraction, as far as we know, and it contracts to zero volume.

As the star contracts, the intensity of its gravitational field at considerable distances does not change, but its surface gravity increases without limit. One consequence of this is that the escape velocity from the star increases steadily as the star contracts. It becomes harder and harder for any object to free itself and get away from the star as that star contracts and its surface gravity increases.

At the present moment, for instance, the escape velocity from the surface of our sun is 617 kilometers (383 miles) per second, nearly 55 times the escape velocity from the surface of the Earth. That is still small enough for material to escape from the sun rather easily. The sun (and other stars) is constantly emitting subatomic particles in all directions at high speed.

If the sun were to contract, however, and if its surface gravity were to increase, its escape velocity would increase into thousands of kilometers per second—tens of thousands—hundreds of thousands. Eventually, the escape velocity would reach a figure of 300,000 kilometers (186,000 miles) per second, and that is the speed of light.

When the star (or any object) contracts to the point where the escape velocity equals the speed of light, it has reached the "Schwarzschild radius," so-called because it was first discussed by the German astronomer Karl Schwarzschild (1873–1916), though the first full theoretical treatment of the situation was not advanced till 1939 by the American physicist J. Robert Oppenheimer (1904–67).

The Earth would reach its Schwarzschild radius if it were to shrink to a radius of 1 centimeter (0.4 inches.) Since the radius of any sphere is half its diameter, the Earth would then be a ball 2 centimeters (0.8 inches) across, a ball that would contain the entire mass of the Earth. The sun would reach its Schwarzschild radius if it were to shrink to a radius of 3 kilometers (1.9 miles) while retaining all its mass.

It is well established that nothing with mass can travel at more than the speed of light. Once any object shrinks to its Schwarzschild radius or less, then nothing can escape from it.* Anything that falls into the contracted object can't get out again, so that the contracted object is like an infinitely deep hole in space. Even light can't get out so that the contracted object is utterly black. The American physicist John Archibald Wheeler (1911–) was the first to apply the term "black holes" to such objects.†

* This is not *quite* true, it has recently turned out. I'll explain that later on.
† Oddly enough, the French astronomer Pierre Simon de Laplace (1749–1827) speculated on the possibility of objects so massive that nothing could escape, not even light, as long ago as 1798.

It would seem, then, that black holes are bound to form when stars run out of fuel and are large enough to produce a gravitational field sufficient to contract it to its Schwarzschild radius. This would seem to be a one-way process. That is, a black hole can form but it cannot unform again. Once it has formed, it is—barring an exception I will discuss later—permanent.

Furthermore, anything approaching a black hole is likely to be captured by the enormously intense gravitational field that exists in its near vicinity. The approaching object may spiral about the black hole and, eventually, fall into it. Once that happens, it can never emerge. It would seem, therefore, that a black hole can gain mass, but cannot lose it.

If black holes form, then, but never vanish, there must be a steady increase in the number of black holes as the universe ages. Furthermore, if each black hole can add to its mass, but not diminish, all the black holes must be constantly growing. With more and bigger black holes each year, a larger and larger percentage of the mass of the universe is to be found in black holes as time goes on, and, eventually, every object in the universe will find itself in one black hole or another.

If we live in an open universe, then, we might imagine that the end is not just maximum entropy and heat-death in an endless sea of thin gas. It is not even maximum entropy and heat-death in each of a billion galactic clusters separated, each from all the others, by incalculable and ever-growing distances. Instead, it would seem that the universe would, in the far future, attain maximum entropy in the form of a number of enormously massive black holes, existing in clusters that would each be separated from all the others by incalculable and ever-growing distances. This, indeed, would seem to be, right now, the most likely future for an open universe.

There are theoretical reasons for supposing that vast quantities of work can be done by the gravitational energies of black holes. We can easily imagine human beings using black holes as a universal furnace, throwing in unneeded mass and making use of the radiation produced in the process. If no surplus mass existed, it might be possible to make use of the rotational energy of a black hole. In this way, much more energy can be extracted from black holes than from the same mass of ordinary stars, and the human species might last longer in a universe with black holes than in one without them.

In the end, however, the second law will have its way. All matter would have ended in black holes, and the black holes would no longer be rotating. No further work could then be extracted from them and maximum entropy would exist. It seems that it would be far harder to evade the heat-death with black holes than without, once that heat-death comes. Random fluctuations into patches of low entropy could not be easily envisaged if we are dealing with black holes and it is difficult to see how life could then avoid its final catastrophe.

How would black holes fit in with a closed universe, however?

The process whereby black holes increase in number and in size may be a slow one considering the total size and mass of the universe. Although the universe is now 15 billion years old, black holes probably still make up only a small portion of its mass.* Even after half a trillion additional years, when the turnabout comes and the universe begins to contract, the black holes may still only make up a small fraction of the total mass.

Once the Universe starts contracting, however, the black-hole catastrophe gains additional potential. The black holes that formed during the period of expansion were in all likelihood confined to the cores of galaxies, but now as the galactic clusters approach each other and as the universe grows richer and richer in energetic radiation, we can be sure that black holes will form in greater numbers and grow more quickly. In the final stages, as the galactic clusters coalesce, the black holes coalesce, too, and the ultimate compression into a cosmic egg is certainly a compression into an enormous universal black hole. Nothing with the mass of the entire universe and the dimensions of the cosmic egg could possibly be anything but a black hole.

But, then, if nothing can emerge from a black hole, how can the cosmic egg formed by the contraction of the universe explode to form a new universe? For that matter, how could the cosmic egg which existed 15 billion years ago have exploded to form the universe we now inhabit?

To see how that can be, we must realize that not all black holes are equally dense. The more mass an object has, the more intense its surface gravity is to begin with (if it is an ordinary star) and the higher its escape velocity. The less it need contract, therefore, to raise the escape velocity to a value equal to the speed of light and the larger the Schwarzschild radius it ends with.

---

* We can't be quite sure of that. Black holes are almost impossible to detect and there may conceivably be many existing that escape our notice. It may even be that it is the mass of these unnoticed black holes that represent the "missing mass" needed to make our universe a closed one—in which case black holes could make up anywhere from 50 to 90 percent of the mass of the universe.

As I said earlier, the sun's Schwarzschild radius would be 3 kilometers (1.9 miles). If a star with a mass of three times that of the sun were to contract to its Schwarzschild radius, that radius would be equal to 9 kilometers (5.6 miles).

A sphere with a radius of 9 kilometers would have 3 times the radius of a sphere with a radius of 3 kilometers and would have 3 × 3 × 3, or 27 times the volume. In the 27-fold volume of the larger sphere there would be 3 times the mass. The density of the larger black hole would be only 3/27 or 1/9 the density of the small one.

In general, the more massive a black hole, the less dense it is.

If the entire Milky Way Galaxy, which has a mass about 150 billion times that of the sun, were to contract to a black hole, its Schwarzschild radius would be 450 billion kilometers, or about 1/20 of a light-year. Such a black hole would have an average density of only about 1/1000 that of the air about us. It would seem like a pretty good vacuum to us, but it would still be a black hole from which nothing could escape.

If there were enough mass in the universe to make it closed, and if all that mass were compressed into a black hole, the Schwarzschild radius of that black hole would be about 300 billion light-years! Such a black hole would be far greater in volume than the entire known universe, and its density would be considerably less than the average density of the universe is considered to be at present.

In that case, let us imagine the universe contracting. Each galaxy, let us suppose, has lost most of its matter to a black hole, so that the contracting universe consists of a hundred billion black holes or more, each anywhere from 1/500 of a light-year to 1 light-year in diameter, depending on its mass. No matter can emerge from any of those black holes to any significant degree.

But now, in the final stages of contraction, all those black holes meet and coalesce to form a single black hole with the mass of the universe—and the Schwarzschild radius at a distance of 300 billion light-years. Nothing can get outside that radius, but it may well be that there can be expansions *within* the radius. The swooping outward, so to speak, of that radius may, in fact, be the very event that ignites the big bang.

Once again the universe as we know it forms, expanding outward in a vast explosion. Eventually the galaxies, stars, and planets are formed. Sooner or later black holes with masses the size of stars begin to form and the whole thing starts all over again.

If we argue along these lines, it would seem we must come to the conclusion that the universe cannot be open; that it cannot expand forever.

The cosmic egg out of which the expansion began *must* have been a black hole and it *must* have had a Schwarzschild radius. If the universe were to expand indefinitely then parts of it would have to move outside the Schwarzschild radius eventually, and that would seem to be impossible. Hence, the universe must be closed and the turnabout must come before the Schwarzschild radius is reached.*

Quasars

Of the three catastrophes of the first class that could serve to make life in the entire universe impossible—expansion to heat-death, contraction to cosmic egg, and contraction to separate black holes—the third is different from the first two in important ways.

Both the general expansion of the universe to the heat-death, or its general contraction to the cosmic egg would affect the entire universe more or less equally. In either case, assuming human life still survives a trillion years from now, there would be no reason to suppose that we would be getting a particularly bad break—or a particularly good one—through our position in the universe. Our portion of the universe won't get it in the neck significantly sooner—or later—than any other.

In the case of the third catastrophe, that of separate black holes, the situation is quite different. We are here dealing with a series of *local* catastrophes. A black hole can form here, and not there, so that life may become impossible here, but not there. In the long run, to be sure, everything will coalesce into a black hole, but the black holes that form here and now can make life impossible in their vicinity *here and now*, even though life elsewhere can go on, uncaring and unheeding, for a trillion years. Therefore, we must now ask whether there are indeed black holes now in existence. If there are, we must ask where they are likely to be and how likely it is that any of them will interfere with us catastrophically before (even perhaps long before) the final catastrophe.

To begin with, it stands to reason that a black hole is most likely to form in places where the most mass is already gathered together. The more massive a star, the more likely it is as a candidate for an even-

* This is why, as I said in the previous chapter, I am convinced the universe is closed despite the current balance of evidence in favor of its being open.

tual black hole. Clusters of stars, where numerous stars are crowded together closely, are even better candidates.

The largest, most thickly star-strewn clusters of all are at the centers of galaxies, particularly at the center of giant galaxies like our own or larger. There, millions to billions of stars are packed into a tiny volume, and there the black hole catastrophe is most likely to take place.

As little as twenty years ago, astronomers hadn't the slightest notion that galactic centers were places where violent events took place. The stars were closely spaced in such centers, but even at the center of a large galaxy stars would be separated by perhaps a tenth of a light-year on the average and there would still be room for them to move about without seriously interfering with each other.

If our sun were located in such a region, we would see over 2½ billion naked-eye stars in the sky, of which 10 million would be of the first magnitude or better—but each would be visible as only a dot of light. The light and heat delivered by all those stars might be as much as a quarter of that delivered by the sun and this additional light and heat might make the Earth uninhabitable, but it would be inhabitable if it were farther from the sun; say in the position that Mars is. We might have argued in this way as recently as 1960, for instance, and even have wished the sun *were* located at the galactic center so that we might enjoy so magnificent a night sky.

If we could detect only the visible light coming from the stars, we might never have had cause to change our minds. In 1931, however, the American radio engineer Karl Guthe Jansky (1905–50) first detected radio waves, with wavelengths a million times longer than those of visible light, coming from particular areas in the sky. After World War II, astronomers developed methods for detecting these radio waves, particularly a comparatively short-wave variety called microwaves. Various microwave sources were pinpointed in the sky by the rapidly improving radio telescopes of the 1950s. Several of them seemed to be associated with what seemed to be very dim stars of our own galaxy. Close examination of these stars made it appear, however, that not only were they unusual in emitting quantities of microwaves, but also in that they seemed to be associated with very faint clouds, or nebulosities, surrounding them. The brightest of them, listed in the catalogues as 3C273, showed signs of a tiny jet of matter emerging from it.

These microwave-emitting objects, astronomers began to suspect, were not ordinary stars, though they appeared starlike. They came to be referred to as quasistellar ("starlike") radio sources. In 1964, the Chinese-American astronomer Hong-Yee Chiu shortened the first part of this phrase to "quasar" and these starlike microwave-emitting objects have been known by that name ever since.

The spectra of the quasars were studied but the dark lines that were found could not be identified until 1963. In that year, the Dutch-American astronomer Maarten Schmidt (1929–) recognized that the lines were the kind that were usually present far in the ultraviolet; that is, that they represented light-waves far shorter than the shortest that would affect our retina and that we could see. They existed in the visible region of the spectra of the quasars only because they had been subjected to an enormous red-shift.

That meant the quasars were receding from us at a more rapid rate than any galaxy that could be seen and were, therefore, farther from us than any galaxy that could be seen. The quasar 3C273 is the closest to us and it is over a billion light-years away. Other, more distant, quasars have been discovered by the dozens. The farthest are up to 12 billion light-years away.

To be visible at all at such enormous distances, quasars must be a hundred times as bright as a galaxy such as ours. If they are, it cannot be because they are a hundred times as large as the Milky Way Galaxy and possess a hundred times as many stars as it does. If the quasars were that large, then even at their enormous distances, our large telescopes would reveal them as cloudy patches and not merely as bright dots of light. They must be much smaller than galaxies.

The smallness of the quasars is also shown by the fact that they vary in brightness from year to year; in some cases from month to month. This can't happen in a large body the size of a galaxy. Parts of a galaxy may grow dimmer and parts may grow brighter, but the average is likely to stay the same. For all of it to brighten or dim, over and over, there must be some effect that is felt by all parts of it. Such an effect, whatever it is, must travel from one end of the galaxy to the other and it cannot travel at more than the speed of light. In the case of the Milky Way Galaxy, for instance, it would take any effect at least a hundred thousand years to travel from end to end, and if our galaxy were to brighten and dim as a whole, over and over, we would expect the period of that change in brightness to be a hundred thousand years long or more.

The rapid changes in the quasars showed that they couldn't be more than a light-year in diameter and yet they emitted radiation at rates a hundred times that of our galaxy which is 100,000 light-years in diameter. How could that possibly be? The beginning of an answer may have come as long before as 1943, when a graduate student in astronomy, Carl Seyfert, detected a peculiar galaxy, a member of a group which are now called "Seyfert galaxies."

Seyfert galaxies are not of unusual size or at unusual distances but they have very compact and bright centers that seem unusually hot and active—rather quasarlike, in fact. These bright centers show variations in radiation, as quasars do, and they may also be not more than a light-year in diameter.

If we imagine a very distant Seyfert galaxy with a particularly luminous center, then all we would see would be that luminous center; the rest would be too dim to be made out. In short, it looks very much as though quasars are very distant Seyfert galaxies and that we see only the luminous centers (though the faint nebulosities around the nearer quasars may be a bit of the galaxies showing up). There may be a billion ordinary galaxies for every huge Seyfert galaxy at distances of over a billion light-years, but we don't see the ordinary galaxies. No part of them is quite bright enough to make out.

Galaxies that are not Seyferts also seem to have active centers; centers that in one way or another are sources of radiation, or that give signs of having suffered explosions, or both.

Can it be that the crowding of stars at galactic centers is bound to set off conditions that produce black holes and that the black holes are constantly growing and can be enormous, and that it is these that produce the activity at galactic centers that is responsible for the brightness of the centers of Seyfert galaxies and of quasars?

The question, of course, arises as to how black holes can be the source of the extremely energetic radiation at galactic centers, when nothing can emerge from a black hole, not even radiation. The point is that the radiation need not come from the black hole itself. When matter spirals into a black hole, its extremely rapid orbiting under the lash of the enormously intense gravitational field in the immediate neighborhood of the black hole, causes the emission of intensely energetic radiation. X-rays, which are like light but which have waves only 1/500,000 as long, are emitted in large quantities.

The amount of radiation emitted in this fashion depends on two things—first, the mass of the black hole, since a more massive black hole can engulf more matter more rapidly and produce more radiation in this way; second, the amount of matter in the neighborhood of the black hole. Matter in the neighborhood collects about the black hole and settles into an orbit called an "accretion disc." The more matter in the neighborhood, the larger the accretion disc is apt to be, the larger the quantity of matter spiraling into the black hole itself, and the more intense the radiation produced. A galactic center is not only an ideal site for the formation of a black hole, but it offers nearby matter in maximum amounts. No wonder, then, that there are compact radiation sources at the centers of so many galaxies and why, in some cases, the radiation is so intense.

Some astronomers speculate that every galaxy has a black hole at its center. In fact, it may be that as gas clouds contract, not long after the big bang, the densest portions condense into black holes. Other contractions then take place within gas-regions attracted by the black hole and orbiting around it. In this way, a galaxy would form as a kind of super-accretion disc about a central black hole which would then be the oldest part of the galaxy.

In most cases the black holes would be rather small and would not produce enough radiation for our instruments to detect anything unusual at the center. On the other hand, some black holes may be so enormous that the accretion disks in their immediate neighborhood are made up of intact stars that virtually jostle each other in orbit, and that are eventually swallowed whole—all of it making the regions in the immediate neighborhood of the black hole extraordinarily luminous, and blazing with energetic radiation.

What's more, matter tumbling into a black hole can release up to 10 percent, or even more, of its mass in the form of energy, whereas ordinary radiation from ordinary stars through fusion at the center is the result of the conversion of only 0.7 percent of mass into energy.

Under these conditions, it's not surprising that quasars are so small and yet so luminous. One can also understand why the quasars dim and brighten as they do. That would depend on the irregular manner in which matter would happen to spiral inward. Unusually large clumps of it might enter at some times, rather small quantities at another.

According to studies in 1978 on X-ray radiation from space, it is considered possible that a typical Seyfert galaxy contains central black holes with masses of from 10 to 100 million times that of the sun. The black holes at the center of quasars must be considerably larger still, with masses a billion times that of the sun, or more.

Even galaxies that are not Seyferts can be unusual in this respect if they are large enough. There is a galaxy known as M87, for instance, which is perhaps 100 times the mass of our own Milky Way Galaxy and contains perhaps 30 trillion stars. It is part of a huge galactic cluster in the constellation Virgo, and is 65 million light-years away. Galaxy M87 has a very active center that is less (perhaps much less) than 300 light-years across, as compared to a total diameter of 300,000 light-years for the entire galaxy. What's more, there seems to be a jet of matter spurting out of the center past the galactic limits.

In 1978, astronomers reported on a study of the brightness of the core as compared with its outer regions and on the rate at which stars seemed to be moving near the center of the galaxy. The results of these studies led to the suggestion that there is a huge black hole in the center of that galaxy, one with a mass equal to 6 billion times that of the sun. Enormous as it is, though, that black hole is still only 1/2500 of the mass of galaxy M87.

Within Our Galaxy

Clearly, the black hole at the center of galaxy M87 and the black holes at the centers of the Seyfert galaxies and of the quasars can't very well be dangers to us. The 65 million light-years that separate us from the M87 black hole and the still greater distances that separate us from Seyfert galaxies and from quasars are more than enough insulation against the worst the black holes can do right now. Furthermore, the quasars are all receding from us at enormous velocities, anywhere from one-tenth to nine-tenths the speed of light, and even galaxy M87 is receding from us at a respectable velocity.

In fact, since the universe is expanding, all black holes located anywhere outside our local group are being carried rapidly and steadily away from us. They can in no way affect us until late in the period of contraction, which will then itself serve as the ultimate catastrophe.

But then what about the galaxies of our own local group, which will remain in our vicinity no matter how long the universe continues to expand? Might the galaxies of our local group contain black holes? They might. None of the galaxies of the local group outside our galaxy show any signs of suspicious activity at the centers, and the small members are not likely to have big black holes anyway. The Andromeda galaxy, which is somewhat larger than our own Milky Way Galaxy, might well have a fairly large black hole at its center, and it certainly isn't going to recede from us very much at any time. On the other hand, it isn't going to approach us very much either.

What, then, about our own galaxy? There is suspicious activity at its center. The Milky Way is not really an active galaxy in the sense of M87 or the Seyferts and quasars, but its center is far closer to us than is the center of any other galaxy in the universe. Whereas the nearest quasar is 1 billion light-years away, and M87 is 65,000,000 light-years away, and the Andromeda galaxy is 2,300,000 light-years away, the center of our own galaxy is only 32,000 light-years away. Naturally, we could detect a small activity more readily in our own galaxy than in any other.

The activity of a 40 light-year-wide object at the very center of our own galaxy is large enough to allow the possibility of a black hole. Some astronomers, in fact, are willing to estimate a black hole with a mass as great as 100 million times the mass of our sun to be sitting in the center of our galaxy.

Such a black hole is only 1/60 the mass of the black hole thought to be at the center of galaxy M87, but then our galaxy is far less massive than galaxy 87. Our black hole would have about 1/1500 the mass of our galaxy. In proportion to the size of the galaxy containing it our black hole would be 1.6 times as great as that of M87.

Does the black hole at the center of our Milky Way Galaxy pose a threat to us? If so, how immediate?

We might argue it out this way. Our galaxy was formed soon after the big bang and the black hole at the center may have been formed even before the rest of the galaxy was. Let us say that the black hole was formed 1 billion years after the big bang, or 14 billion years ago. In that case, it took the black hole 14 billion years to swallow up 1/1500 of our galaxy. At that rate it will take some 21,000 billion years to swallow the entire galaxy, by which time either the heat-death catastrophe, or, more likely (I think), the next cosmic-egg catastrophe would have overtaken us anyway.

Is it fair, though, to say "at that rate"? After all, the larger a black hole grows, the more wholesale its engulfing of surrounding matter. It might take 14 billion years to swallow up 1/1500 of our galaxy and only 1 billion years to complete the job.

On the other hand, the ability of a black hole to engulf matter depends also on the density of matter in the neighborhood. As the black hole at the center of any galaxy grows, it will efficiently clean out the stars in the galactic nucleus and will eventually form what we might call a "hollow galaxy," one with a nucleus that is empty except for the giant black hole at the center, one with a mass of up to 100 billion times that of our sun, or even a trillion times in a really large galaxy. Such enormous black holes would be between 0.1 and 1 light-year in diameter.

Even so, the remaining stars in the outskirts of the galaxy would then be orbiting around that central black hole in comparative safety. Every once in a while, a particular star under the influence of other stars might find its orbit twisted in such a way that it would approach uncomfortably close to the black hole and be captured, but that would be a rare incident and with time it would grow rarer. For the most part there would be no more danger in circling the central black hole then there is for the Earth to be circling the sun. After all, if the Earth for any reason approached the sun too closely, it would swallow it up as efficiently as a black hole would.

In fact, even if the black hole at the center of the galaxy cleaned out the nucleus and left the galaxy hollow, we wouldn't be able to tell except for the decline in radiant activity as less and less material spiraled into the black hole. The center of the Galaxy is hidden behind vast dust clouds and star clusters in the direction of the constellation Sagittarius, and if it were emptied we couldn't see any change.

If the universe were an open one, we might picture the far-future expansion as one, perhaps, in which all the galaxies are hollow, a series of super black holes, each surrounded by a sort of asteroid belt of stars working their way toward the heat-death.

Is it possible, though, that there may be black holes in our galaxy elsewhere than at the center and, therefore, closer to ourselves?

Consider the globular clusters. These are tightly packed, spherical groups of stars, the whole being about 100 light-years in diameter. Within that relatively small volume there may be anywhere from 100,000 to 1 million stars. A globular cluster is rather like a detached portion of the galactic nucleus, much smaller than the nucleus, of course, and not so tightly packed. Astronomers have detected something over a hundred of these distributed in a spherical halo about the galactic center. (Undoubtedly, other galaxies also have their halo of globular clusters.)

Astronomers have detected X-ray activity at the center of a number of these clusters and it isn't at all hard to suppose that the same processes that gave rise to black holes at the center of the galaxies would also give rise to black holes at the center of the globular clusters.

The cluster black holes would not be as large as those at galactic centers, but they could be 1,000 times as massive as our sun. Though smaller than the great galactic black hole, could they pose a more immediate danger? At the present moment, certainly not. The nearest globular cluster to us is Omega Centauri, which is 22,000 light-years away, still a safe insulating distance.

So far, then, the breaks seem to be with us. Astronomical discoveries since 1963 have shown the centers of galaxies and of globular clusters to be active, violent places inimical to life. They are places where the catastrophe has already come in the sense that life on any planets in such areas would be destroyed either directly by absorption into a black hole, or indirectly by the deadly bath of radiation resulting from such activity. We might, however, rather say there was never anything to suffer a catastrophe there, since it is unlikely that under such conditions life would have formed in the first place. We ourselves, however, exist in the quiet outskirts of a galaxy where the stars are sparsely strewn. Therefore the black hole catastrophe is not for us.

But wait! Is it possible that even here in the outskirts of the galaxy there are black holes? There are no large clusters in our neighborhood within which black holes can form, but there might be enough mass concentrated into single stars to form a black hole. We must ask, then, if any giant stars near us have formed black holes. If so, where are they? Can we recognize them? Are they a danger?

There seems to be a frustrating fatality about black holes. It is not the black hole that we see directly, but the radiational "death-cry" of matter falling into it. The death-cry is loud when a black hole is surrounded by matter that it can capture, but then the surrounding matter hides the immediate vicinity of the black hole from view. If there is little matter surrounding the black hole so that we have a chance to see the immediate neighborhood, there is also little matter falling into it and the death-cry is weak, so that we are very likely to overlook the black hole's existence.

There is one convenient possibility, however. About half the stars in the universe seem to exist in pairs ("binary systems") revolving about each other. If both are large stars, then one might be converted into a black hole at some particular stage of its evolution and matter from the companion star might, little by little, be drawn into the nearby black hole. That would produce the radiation without obscuring the black hole unduly.

In order to detect possible situations of this sort, astronomers have scanned the sky for X-ray sources and then tried to pin each one down, looking for one that was nearby and that could not be explained by anything less than a black hole. For instance, an X-ray source that changed its intensity in irregular fashion was more likely to be a black hole than one whose intensity was steady, or changed in a regular fashion.

In 1969 an X-ray-detecting satellite was launched from the coast of Kenya on the fifth anniversary of Kenyan independence. It was named Uhuru from the Swahili word for "freedom." It could search for X-ray sources from its orbit beyond Earth's atmosphere—which was necessary, for the atmosphere absorbs X-rays and allows none to reach any waiting X-ray-detecting device on Earth's surface.

Uhuru detected 161 X-ray sources, half of them in our own galaxy. In 1971, Uhuru observed a bright X-ray source in the constellation of Cygnus the Swan—"Cygnus X-1" it was called—and detected an irregular change in intensity. Attention was eagerly focused on Cygnus X-1, and microwave radiation was also detected. The microwaves made it possible to pinpoint the source very accurately and it was found to be just next to, but not on, a visible star. The star was HD-226868, a large, hot, blue star about 30 times as massive as our sun. The star was clearly circling in an orbit with a period of 5.6 days —an orbit the nature of which made it appear that the other star was perhaps 5 to 8 times as massive as our sun.*

The companion star cannot be seen, even though it is a source of intense X-rays, which, considering its mass and the brightness it therefore would have to have, would not be the case if it were a normal star. It must therefore be a collapsed star and it is too massive to have collapsed to anything less than a black hole. If so, it is far smaller than the black holes we have earlier discussed, the ones that are thousands, millions, even billions of times as massive as our sun. This one is at most only 8 times the mass of our sun.

It is, however, closer than any of those others. Astronomers estimate that Cygnus X-1 is only 10,000 light-years away from us, less than a third the distance of the galactic center, and less than half the distance of the nearest globular cluster.

In 1978, a similar binary system was reported in the constellation of Scorpio. The X-ray source there, listed as V861Sco, may represent a black hole with a mass as much as 12 times that of the Sun, and it is only 5,000 light-years away.

We can argue correctly that even 5,000 light-years is an adequate

---

* It isn't easy to determine the mass of a star all by itself. However, if a pair of stars are circling each other, their masses can be determined from the distance between them and the length of time it takes them to complete the circle, as well as the location of the center of gravity between them.

insulating distance. We can further argue that it is rather unlikely that there are black holes much closer than that. The kind of stars that produce black holes are so few that it is not likely that one of them would happen to be close to us under conditions where we would remain unaware of it. If it were close enough, even minor amounts of matter falling into it would produce detectable intensities of X-rays.

These nearby black holes, however, have a danger that others do not. Consider: All black holes in galaxies outside our local group are particularly far away and are constantly moving farther away because of the expansion of the universe. All black holes in galaxies other than ours but inside the local group are still far away and, on the whole, maintain their distance. Though they do not move appreciably farther from us, neither do they move appreciably nearer. The black hole at the center of our galaxy is, of course, closer to us than any black hole in any other galaxy, but it, too, maintains its distance, for the sun moves about it in a nearly circular orbit.

The black holes in our galaxy that are not at the center, however, all move as we do about the center of the galaxy. We all have our orbits and in the course of moving about them, those black holes may recede from us or may approach us. Half the time, in fact, they are bound to approach us.

How closely? How dangerously?

It is time then to pass from catastrophes of the first class that affect the universe generally, to catastrophes of the second class that affect our solar system particularly.

Part II

Catastrophes of the Second Class

Chapter 5

Collisions
with the Sun

Birth by Close Encounter

It would seem that the most likely and the most nearly unavoidable catastrophe of the first class is the coming of the next cosmic egg, perhaps a trillion years from now. The discussion of black holes, however, has shown that local catastrophes could strike particular places long before the trillion-year period is up. It is time, then, to consider the chance of a local catastrophe rendering our solar system uninhabitable and thus putting an end to human life, even while the rest of the universe remains untouched.

This would be a catastrophe of the second class.

Before the time of Copernicus, it seemed self-evident that the Earth was the motionless center of the universe, with all else revolving about it. The stars, in particular, were considered to be fixed to the outermost sphere of the sky and to revolve in one piece, so to speak, about the Earth in twenty-four hours. The stars were referred to as "fixed stars" to differentiate them from those nearer bodies—the sun, the moon, the planets—which revolved independently.

Even after the Copernican system removed the Earth from its central position, that did not at first affect the view of the stars. They still seemed bright, immovable objects fixed to an outermost sphere, while within that sphere the sun was at the center and the various planets, including the Earth, circled it.

In 1718, however, the English astronomer Edmund Halley (1656–1742), recording the position of the stars, noted that at least three stars—Sirius, Procyon, and Arcturus—were not in the spots recorded by the Greeks. The difference was substantial and the Greeks could not have made so large a mistake. It seemed clear to Halley that these stars had moved relative to the others. Since then, more and more stars have shown such a "proper motion" as astronomers' instruments for detecting such motion have grown more delicate.

Clearly, if various stars move through space at equal rates, the change in position of a very distant star would be far less to our observation than that of a fairly close star. (We know from experience how slowly a distant airplane seems to move compared with one that is much closer.) The stars are so distant that only the closest can show a detectable proper motion, but from it, it seems a fair conclusion that all stars move.

To be sure, the proper motion of a star is only its motion across our line of sight. A star could also be moving toward us or away from us, and that part of its motion would not show up as proper motion. In fact, it could be moving directly toward us or directly away from us so that there would be no motion at all across the line of sight even though it might be comparatively near to us.

Fortunately, by means of the Doppler-Fizeau effect, described earlier, the speed of approach or recession can also be determined and the three-dimensional "space velocity" of at least the nearer stars can be worked out.

And why should not the sun be moving, too, then?

In 1783, the German-British astronomer William Herschel (1738–1822) studied the proper motions that were by then known. It seemed that the stars in one half the sky tended, on the whole, to be moving apart from each other. In the other half, they tended to be moving together. Herschel decided that the most logical way of explaining this was to suppose that the sun was moving in one particular direction toward the constellation Hercules. The stars we were approach-

ing seemed to be moving apart as we approached, and the stars behind us seemed to be closing together.

When astronomical objects move through space, it is quite likely
that one will move about another, if they are sufficiently close to each
other so that they are intensely affected by each other's gravitational
field. Thus, the moon circles the Earth, while the Earth and the other
planets, move around the sun. Again, one star in a binary system will
move about another.

Where the objects are all far from each other, however, and when
there is no one object that by its enormous mass predominates over
all the others (as the sun predominates over all the smaller bodies of
the solar system) the motions are not a simple circling of one object
about another. Instead, there would seem to be an almost random
motion, like that of bees in a swarm. Through the nineteenth century,
it seemed that such a bees-in-a-swarm motion characterized the stars
about us, and at that time, it didn't seem illogical to suppose that in
these random motions one star might just happen to jostle another.

In fact, in 1880, the English astronomer Alexander William Bick-
erton (1842–1929) suggested that that might be how the solar system
had come into being. A long time ago, he thought, a star had passed
by the sun and by the gravitational effect of each upon the other,
material was pulled out of both which later condensed into planets.
The two stars had approached as single bodies and had left, each one,
with the beginnings of a planetary system. It was a rather dramatic
example of what could only be described as a cosmic rape. This
"catastrophic theory" of solar system origin was more or less accepted by astronomers, with a variety of modifications, for over half
a century.

It is clear that while such a catastrophe might mark the beginning
of the world for us, it would, if repeated, mark the catastrophic end
of it. Another close approach of a star to our sun would, for a long
time, subject us to the increasing heat of an approaching second
luminary, while our own sun would be destabilized in one fashion or
another by the increasing gravitational effect upon us. That same
effect would produce increasingly serious disturbances in Earth's
orbit. It seems very unlikely that life could withstand the enormous
effects of this on conditions on Earth's surface.

How likely, then, is it that such a near-collision will take place?

Not very likely at all. In fact, one of the reasons why the catastrophic theory of solar system origin did not, in the end, survive was
that it involved such an unlikely event. In the outskirts of the Galaxy,

where we are located, stars are so far apart, and move so slowly compared to the huge distances of separation that collisions are difficult indeed to imagine.

Consider Alpha Centauri, which is the star closest to us.* It is 4.4 light-years away from us and approaching. It isn't approaching us squarely, for it is also moving sideways. The result is that it will eventually be about 3 light-years from us, at which point it will pass us (without being close enough to affect us in any significant way) and begin to recede.

Suppose, however, it were approaching us squarely. Alpha Centauri is moving through space, relative to us, at a speed of 37 kilometers (23 miles) per second. If it were aiming at us directly at this speed, it would pass through our solar system in 35,000 years.

On the other hand, suppose Alpha Centauri was aimed only 15 minutes of arc away from an actual collision with the sun, a miss that would represent half the width of the full moon as it appears to us. This would be like supposing that we were trying to hit something dead-center on the face of the moon, but missed and hit the rim of the moon instead. If Alpha Centauri's aim were no better than that, it would miss us by 1/50 of a light-year or about 180 billion kilometers (110 billion miles). This would be thirty times the distance of Pluto from the sun. Alpha Centauri would then be an extraordinarily bright star in the sky but its effect on Earth from this distance would be negligible.

Another way of looking at it is this. The average separation between stars in our part of the Galaxy is 7.6 light-years and the average velocity at which they are moving relative to each other is perhaps 100 kilometers (62 miles) per second.

Let us reduce light-years to kilometers and imagine the stars (reduced in proportion) to be 1/10 of a millimeter across. These tiny stars, which would resemble small pieces of grit just visible to the eye would be distributed at an average separation of 7.6 kilometers (4.7 miles). If viewed on a two-dimensional field, there would be fourteen of them scattered over the area of the five boroughs of New York City.

Each would be moving at a speed (reduced in proportion) of 30 centimeters (1 foot) a year. Imagine, then, these fourteen pieces of

---

* Actually, it is a binary star, two stars circling each other, with a third dwarf star comparatively far off from those two. Among the stars in our neighborhood we can even find six stars, three binary pairs, bound to one another gravitationally. For our purposes here, I will use the word "star" to include star-systems of from two to six stars that are gravitationally bound.

grit scattered over the five boroughs and each moving 1 foot a year in random directions, and ask yourself what are the chances that two of them will eventually collide.

It has been estimated that, in the outskirts of the Galaxy, the chances of a near approach of any two stars is not more than 1 in 5 million over the entire 15-billion-year lifetime of the Galaxy. This means that even in the trillion years before the next cosmic egg, there is only 1 chance in 80,000 of a near approach of a star to our own. This type of catastrophe of the second class is so much less likely than any catastrophe of the first class, that it seems unnecessary to worry about it at all.

Then, too, the possible collision-approach of a star, given our present level of astronomic expertise (let alone the higher levels that may be developed in the future) would give us warning many thousands of years in advance. Catastrophes, when they come, are much more dangerous if they are sudden and unexpected, leaving us no time to adopt countermeasures. Although a star collision would find us helpless now even if we had had warning many thousands of years ago, this may not necessarily be so in the future (as I shall explain later), and from here on in we might expect that the warning will come in plenty of time for evasion or avoidance.

For both these reasons—the extremely low chance of its happening and the certainty of a very long warning period—it makes no sense to worry about this particular catastrophe.

Mind you, by the way, that it doesn't matter whether the invading star is a black hole or not. The black hole could not kill us more effectively than an ordinary star could, though a large black hole equal in mass to 100 times that of our sun could exert its deadly effect at ten times the distance an ordinary star could manage, so that the accuracy with which it bore down on us would not need to be so fine.

However, it is very likely that large black holes are, at best, so rare, that even allowing for their greater sphere of action, the chance of one of them approaching catastrophically close is millions of times less than the already tiny chance of an ordinary star doing so.

To be sure, there are objects other than stars that might make catastrophic approaches, and those other objects might, in some cases, come with little or no warning—but we'll take up such cases in due course.

Orbiting the Galactic Nucleus

One reason for the unlikelihood of a catastrophic encounter of our sun with another star rests in the fact that the stars in our vicinity are *not*, after all, moving randomly as bees would in a swarm. We might find this random motion in the center of the Galaxy or in the center of a globular cluster, but not out here.

In the outskirts of the Galaxy, the situation is rather like that in the solar system. The galactic nucleus, which takes up a rather small central portion of the Galaxy, has a mass of tens of billions of that of the sun, part of which, of course, could be the central black hole, assuming it exists. This nucleus, acting as a whole, serves as the Galaxy's "sun."

The billions of stars in the galactic outskirts circle the galactic nucleus in orbit, as the planets circle the sun. The sun, for instance, which is 32,000 light-years from the galactic center, is moving about that center in a nearly circular orbit at a velocity of about 250 kilometers (155 miles) per second, and it takes it about 200 million years to complete one revolution. Since the sun was formed nearly 5 billion years ago, this means that it has completed twenty-four or twenty-five turns about the galactic center in its lifetime, assuming that its orbit has been the same in all this time.

Naturally, stars that are closer to the galactic center than is the sun move more rapidly and complete the revolution in less time. As they gain on us, they approach us, but having passed us at, presumably, a safe distance, they then recede from us. In the same way, stars that are farther from the galactic center move less rapidly and complete the revolution in a longer period. While we are overtaking such stars, they seem to approach us, but having passed them at, presumably, a safe distance, they then recede from us.

If all the stars were moving in very nearly circular orbits in very nearly the same plane and at widely different distances from the point about which they revolve (as is true of planets within the solar system) there would be no chance of any collision or near-collision ever. As a matter of fact, in the 15 billion years of the Galaxy's history, the stars seem to have jostled themselves into very much this arrangement so that the outskirts of the Galaxy form a flat ring (within which the stars are arranged in a set of spiral structures) whose plane passes through the center of the galactic nucleus. The fact that the sun has made twenty-five circuits of its orbit without any sign of mishap that we can detect in Earth's geological record shows the efficiency with which this arrangement works.

There are, however, only nine major planets in the solar system, while there are billions of sizable stars in the outskirts of the Galaxy. Even though the majority of stars are orbitally well-behaved, even a small percentage of mavericks means a large number of stars whose orbits are troublesome.

Some stars have orbits that are quite elliptical. It may well be that the orbit of such a star skims ours and is separated from it at some point by a relatively small distance; but that every time that the sun has been at the skim point, the other star has been far away and vice versa. Eventually, it would be inevitable that the sun and the other star should both reach the skim point at nearly the same time and undergo a close approach—but that could be a very long "eventually."

What is worse is that orbits do not necessarily stay the same. When two stars make a moderately close approach, one that isn't nearly close enough to disrupt the planetary systems (if any) of either, the mutual gravitational effect may alter the orbits of both just a bit. Even though the sun may not itself be involved in such an approach, it may be affected. Two other stars may make a close approach on the other side of the Galaxy, for instance, and one of them may have its orbit altered (or "perturbed") in such a way that where previously it had never approached the sun's orbit, it now has the potentiality of approaching the solar system.

It works the other way, too, of course. A star whose orbit might bring it uncomfortably close to the solar system, may, as a result of a perturbation not involving us, shift its orbit so as to come nowhere close to us.

Elliptical orbits present another interesting problem. A star with a markedly elliptical orbit may now be in our portion of the Galaxy, but hundreds of millions of years from now, it may have moved to the other end of its orbit much farther from the galactic nucleus than it is now. Such an elliptical orbit, in which the present position of the star in our neighborhood places it at or near its closest approach to the galactic nucleus, is not a dangerous one. Nothing much can happen to it way out there.

An elliptical orbit can also place a star in our neighborhood at or near the far point of its orbit and a hundred million years from now it can have plunged deeper into the Galaxy and be skimming the galactic nucleus at a much smaller distance. That may conceivably spell trouble.

The stars are more thickly spread the closer one gets to the nucleus and the orbits are less regular and stable. A star moving inward increases its chance of perturbation. Outright collision remains a very small probability, but is substantially greater than in the outskirts. The chance of an approach close enough to introduce an orbital perturbation rises by perhaps the same ratio and becomes large enough to become perceptible.

There may be a good chance that every star on the outskirts whose elliptical orbit will take it closer to the nucleus will emerge with an at least slightly modified orbit, one which, if not dangerous to us before, might become dangerous (or vice versa, of course). In fact, a perturbation could affect us directly.

Earlier I spoke of the case of a star skimming by us at a distance from the sun thirty times the distance of the outermost planet, Pluto. I said it would in no way affect us. It would not, in the sense that it would not seriously affect the workings of the sun or the environment on Earth. All the less so if it passed at a distance of a light-year or so.

And yet some passing star, which is not close enough to cause us the slightest trouble in the way of extra heat, may very slightly slow the sun in its progress about the galactic center. In that case the sun's nearly circular orbit may be made slightly more elliptical and it may swoop in somewhat closer to the galactic nucleus than it has ever done before in its two dozen revolutions.

Closer to the galactic nucleus, the chances for further perturbation become somewhat greater and further changes may take place. Given a streak of bad luck, the sun may finally be in an orbit which will take us so close to the inner region of the galaxy, say a billion years from now, that the general radiation background may be strong enough to wipe out all life. The chances of this are all very small, however, and it may all be included in the 1 chance in 80,000 over the next trillion years.

That 1-in-80,000 chance over the next trillion years involves individual stars, however. What about globular clusters? The globular clusters are not located in the galactic plane but are distributed about the galactic nucleus in a spherical shell. Each globular cluster revolves about the galactic nucleus, but its plane of revolution is inclined to the galactic plane at a large angle. If a globular cluster is now located far above the galactic plane, it will, as it moves along its orbit, come down at a slant, move through the galactic plane, sink far below it, then come up at a slant and move through the galactic plane at the opposite side of the galactic nucleus, and return to where it now is.

If a globular cluster is as far from the galactic nucleus as we are, then every 100 million years or so it will pass through the galactic plane. If it is closer to the nucleus, it will do so at shorter intervals, if farther, at longer intervals. Since there may be up to 200 such clusters altogether, we can expect that, at an average, some globular cluster or other will be moving through the galactic plane every 500,000 years or so, if the average distance of globular clusters from the galactic nucleus is equal to that of the solar system.

A globular cluster has a cross-sectional area that is a billion billion times that of an ordinary star and, in crossing the galactic plane is a billion billion times more likely to collide with some star than would be the case if a single star were to cross the galactic plane.

To be sure, the nature of the collisions is not the same. If our sun were hit by a star it would be a clear case of a collision. If our sun were hit by a globular cluster, on the other hand, there might be no real collision at all. Although the globular cluster seems crowded with stars when viewed from a distance, it is still very largely empty space. If our sun were to pass through a globular cluster at random, the chances would be only one in a trillion that it would strike an individual star in that cluster. (Not much of a chance, but far greater than if the sun were to pass through the galactic outskirts with only other individual stars in the neighborhood, as it is doing.)

Still, even though a globular cluster is not likely to damage the sun physically in case of a collision, or even to affect seriously the Earth's environment through mere light and heat, there would be a fairly respectable chance that the sun's orbit would be changed as a result and, just possibly, not for the better.

The possibility of perturbation would increase as the collision was more and more on-the-nose, so to speak, so that the sun would pass through the globular cluster on a path that would take it nearer and nearer the center of the cluster. Not only are the stars more thickly strewn in the center so that the chance of perturbations and the possibility of actual collision would increase, but the sun might then approach a black hole with a mass of a thousand suns that might lie at the center.

The chance of perturbation, or even capture, might be a serious one, and even if not, the energetic radiation in the neighborhood of a black hole might put an end to life on Earth without affecting the physical structure of the planet at all.

The chances of any of this happening are very small. There are not many globular clusters and only those which pass through the galactic plane within a dozen light-years of Earth's distance from the galactic nucleus can offer us danger. At the best, one or two might do so, and the chances of their passing through the plane just as the sun is approaching that portion of its vast orbit are very small indeed.

Furthermore, the impending collision of a globular cluster with us is even less Damoclean than the close approach of a single star would be. A globular cluster is a far more prominent object than a star is, when both are at the same distance, and if a globular cluster were moving in such a way as to give rise to fears of collision, we would be bound to have a million years or more of warning.

Mini-Black Holes

As far as collisions with visible objects are concerned, we *know* that the sun is safe for millions of years. Nothing visible is heading in our direction from a distance close enough to reach us in that time. Might there not be objects in space that we don't detect and of whose existence we are unaware? Might one of these not be approaching, and even be on a collision course with the sun, giving little or no warning? What about black holes of the size of Cygnus X-1; black holes that are not the gigantic ones at the centers of galaxies and globular clusters, and which remain there, but black holes the size of stars which wander in orbits about the galactic centers? To be sure, Cygnus X-1 reveals its presence by the great quantities of matter it swallows up from its perfectly visible companion star. Suppose, though, a black hole was formed through the collapse of a single star, without companions.

Let us say that such a single-star black hole has a mass five times that of our sun, and a radius, therefore, of 15 kilometers (9.3 miles). There is no companion star whose presence will give it away; no companion star to feed it mass and produce a vast radiation of X-rays. There would be only the thin wisps of gas between the stars to feed it, and that will produce only a tiny sparkle of X-rays which will not be particularly noticeable at any distance.

Such a black hole might be within a light-year of us and be too small physically and too inactive radiationally to detect. It might be heading right for the sun and we would not know. We might not know until it was almost upon us and its gravitational field was introducing some unexpected perturbations in our planetary system, or when a very faint but steadily strengthening X-ray source was detected. We might then have no more than a few years' warning of the end of our world. Even if it passed through the solar system without collision, its gravitational field might wreak havoc with the finely tuned celestial mechanics of the solar system.

Is there any likelihood of this happening? Not really very much. It takes a very large star to collapse into a black hole, and there aren't very many large stars. It may be that at best there is only one star-sized black hole in the Galaxy for every 10,000 visible stars. If there is only 1 chance in 80,000 that an ordinary star will collide with the Sun over the space of a trillion years, there is only 1 chance in 800 million that a star-sized black hole will. It might happen within the next year but the odds are nearly a sextillion to one that it won't and it would be entirely unreasonable to worry about such a possibility.

Part of the reason the odds against catastrophe are so huge is that the number of star-sized black holes is so small. It is well known, however, that among any class of astronomical bodies, the smaller varieties are more numerous than the larger ones. Might there not be small black holes that are much more numerous than large ones? A small black hole might not do as much damage when it strikes as a large one would, but it might do damage enough; and, because the small ones are so numerous, the chances of a strike might grow alarmingly high.

In our universe today, however, it would seem very unlikely to find black holes that are less than several times the mass of the sun. A large star might compress itself into a black hole under the pull of its own gravitational field, but there seem to be no compressional forces available to form a black hole out of anything smaller than a large star.

That does not end the danger, however. In 1974, the English physicist Stephen Hawking suggested that in the course of the big bang, the whirling masses of matter and radiation produced incredible pressures here and there which, in the first moments of the formation of the universe, produced innumerable black holes of all masses from that of a star down to tiny objects of a kilogram or less. The black holes of less-than-stellar masses Hawking called "mini-black holes."

Hawking's calculations showed that black holes do not truly retain all their mass, but that it *is* possible for matter to escape from them. Apparently, it is possible for pairs of subatomic particles to form right at the Schwarzschild radius and to speed off in opposite directions. One of the particles plunges back into the black hole, but the other escapes. This steady escape of subatomic particles causes the black hole to behave as though it has a high temperature and is slowly evaporating.

The less massive a black hole is, the higher its temperature, and the more rapidly it tends to evaporate. This means that as a mini-black hole shrinks through evaporation, its temperature rises, and the rate of evaporation increases steadily, till the last bit of the mini-black hole goes with explosive force and it vanishes.

Very small mini-black holes would not have endured through the 15-billion-year history of the universe and would already have completely disappeared. If a mini-black hole had a mass of more than an iceberg to begin with, however, it would be cool enough and would evaporate slowly enough to be in existence still. If in the course of its lifetime it managed to pick up mass, as it is very likely to have done, it would have cooled further and its lifetime would be further extended.*

Even allowing for the disappearance of the smallest (and most numerous) of the mini-black holes, there may still be very many mini-black holes in existence with masses varying from that of a small asteroid to that of the moon. Hawking has estimated that there may be as many as three hundred mini-black holes per cubic light-year in the Galaxy. If they followed the distribution of matter generally, then most of them are in the galactic nucleus. In the outskirts, where we are, there may be only as many as thirty mini-black holes per cubic light-year. This would mean an average separation between mini-black holes of about five hundred times the distance between the sun and Pluto. The nearest mini-black hole to us is likely to be at a distance of 1.6 trillion kilometers (1 trillion miles).

Even at that distance (very close, by astronomic standards) there is plenty of room for it to maneuver in, and not much likelihood of its doing damage. A mini-black hole must make a direct hit to do damage, whereas a star-sized black hole need not. A star-sized black hole

---

* Black holes as massive as stars have effective temperatures within a millionth of a degree of absolute zero and evaporate so slowly that it would take them trillions of trillions of trillions of times as long as the length of time to the next cosmic egg to evaporate. In the meantime, they would undoubtedly have picked up prodigious quantities of mass. Stellar-sized black holes are therefore permanent objects and grow steadily larger, never smaller. The new views of Hawking show their effect only in mini-black holes, and in particularly small mini-black holes at that.

might miss the sun by a substantial distance, but in passing near the solar system, might produce tidal effects in the sun that could seriously alter its properties. It might also perturb the Sun's orbit significantly, with disadvantageous results; or, for that matter, perturb Earth's orbit disastrously.

A mini-black hole, on the other hand, might pass through the solar system without any noticeable effect whatever, upon either the sun or any of the major planets and satellites. For all we know, any number of mini-black holes have skimmed by us and a few may have moved in among the planets without doing us any harm.

What would happen, though, if a mini-black hole actually struck the sun? As far as its mass is concerned, the chances are that it would have no serious effect on the sun. Even if it had the mass of the moon, that would be only 1/26,000,000 that of the sun, about what a tenth of a drop of water is to you.

Mere mass is not all that counts though. If it were the moon that were heading for a collision with the sun, then, unless the moon were moving very quickly indeed, it would vaporize by the time it struck the sun. Even if part of it remained solid by collision time, it would not penetrate very far before vaporizing.

A mini-black hole, however, would not vaporize or be in any way affected by the sun. It would merely burrow in, absorbing mass as it went, with the production of enormous energies. It would grow as it went and pass all the way through the sun, emerging a considerably larger mini-black hole than it entered.

What the effect on the sun might be is very hard to predict. If the mini-black hole struck a glancing blow and just passed through the upper layers of the sun, the effect might not be very deadly. If the mini-black hole struck the sun squarely, however, and burrowed right through its center, it would distrupt the very region of the sun in which nuclear reactions are taking place and solar energies are being produced.

What would then happen I don't know; it would depend on how quickly the sun could "heal" itself. It is perhaps possible that energy production would be disrupted and that before it could be resumed, the sun would collapse or explode. Either way, if it happened unexpectedly enough and soon enough, it would be the absolute catastrophe for us.

Indeed, suppose the mini-black hole were to strike the sun at a rather low speed relative to the sun. The resistance it would meet in passing through the sun's substance might slow it to the point where it would not leave, but would remain within the sun, settling to its center.

Then what? Would it slowly consume the matter of the sun from within? If so, we might not be able to tell the difference from outside. The sun would retain its mass and its gravitational field unchanged; the planets would continue to circle unperturbed; and the sun might even emit its energies as though nothing were happening. But surely, at some crucial point, there would not be enough normal matter to maintain the sun in its present form. All of it would collapse into a black hole with the emission of a vast quantity of killing radiation which would destroy all life on Earth. Or, even if we could imagine somehow surviving the blast of radiation, Earth would then be circling a black hole with all the mass of the sun (so that Earth's orbit would remain unchanged) but one that was too small to see and gave off no radiation to speak of. Earth's temperature would drop close to absolute zero and that would kill us off.

Could it be that a mini-black hole struck the sun a million years ago and has been at work ever since? Might the sun, totally without warning, collapse at any moment?

We can't answer with an absolute no, but let us remember that even with mini-black holes as numerous as Hawking thinks, the chances of hitting the sun are very small; those of hitting the sun dead center are still smaller; those of hitting the Sun dead center and at a speed relative to the sun so small as to allow the mini-black hole to be captured are still smaller. Then, too, Hawking's figures represent a reasonable maximum. It is quite likely that mini-black holes are rarer than that, perhaps even considerably rarer. That would reduce the chances accordingly.

In fact, there is no evidence for mini-black holes at all, except for Hawking's calculations. No mini-black holes have actually been detected; nor has any phenomenon been detected for which the explanation might involve a mini-black hole. (Even the existence of star-sized black holes such as that represented by Cygnus X-1 depends on evidence that has not yet convinced all astronomers.)

More information about the universe must be obtained before we can work out sensible odds in connection with this kind of catastrophe, but we can still be confident that they are very strongly in favor of noncatastrophe. After all, the sun has been in existence for five billion years without collapsing; nor have we happened to observe any star suddenly winking out as though it had finally been swallowed by a mini-black hole at its center.

Antimatter and Free-Planets

An unaccompanied black hole is not the only object in the universe that could conceivably sneak up on us unnoticed. There is another kind of object that is almost as dangerous, but whose existence is even more problematical.

The ordinary matter about us consists of atoms which are made up of tiny nuclei surrounded by electrons. The nuclei are made up of two types of particles, protons and neutrons, each of which is somewhat more than 1,800 times as massive as electrons. Thus, the matter about us is made up of three types of subatomic particles: electrons, protons, and neutrons.

In 1930, Paul Dirac (who first suggested that gravity might be weakening with time) showed that, in theory, there ought to exist "antiparticles." There ought to be, for instance, a particle like the electron but carrying an opposite electric charge. Whereas the electron carried a negative electric charge, its antiparticle would carry a positive one. Two years later, the American physicist Carl David Anderson (1905–) actually detected this positively charged electron. It was named the "positron" though it can also be referred to as an "antielectron."

In due time, the "antiproton" and the "antineutron" were also discovered. Whereas the proton carries a positive electric charge, the antiproton carries a negative one. The neutron carries no charge and neither does the antineutron, but they are opposite in certain other properties. The antielectron, antiproton and antineutron can come together to form "antiatoms" and these can conglomerate into "antimatter."

As it happens, if an antielectron encounters an electron they will annihilate each other, the properties of one cancelling the opposing properties of the other, and the mass of the two will be converted into energy in the form of "gamma rays." (Gamma rays are like X-rays but have shorter waves and are therefore even more energetic.) In the same way, an antiproton and a proton can annihilate each other and so can an antineutron and a neutron. In general, antimatter can annihilate an equivalent mass of matter, if the two encounter each other.

The amount of energy released in such "mutual annihilation" is tremendous. Hydrogen fusion, such as explodes our hydrogen bombs and powers the stars, converts about 0.7 percent of the fusing matter to energy. Mutual annihilation, however, converts 100 percent of matter to energy. Thus a matter-antimatter bomb would be 140 times as powerful as a hydrogen bomb of the same mass.

It works the other way around. It is possible to convert energy into matter. However, just as it takes a particle and an antiparticle together to produce the energy, so energy when converted to matter always produces both a particle and its corresponding antiparticle. There seems to be no way around this.

In the laboratory, the physicist can manufacture particles and antiparticles just a few at a time, but in the period after the big bang, energy was converted into matter in quantities sufficient to form an entire universe. If that were so, however, antimatter must have been formed in precisely the same quantities. Since this must be so, where is the antimatter?

On the planet Earth, there is only matter. A few antiparticles can be formed in the laboratory, or are present in cosmic rays, but they amount to nothing, and the individual antiparticles disappear almost at once as soon as they encounter the equivalent particles, giving off gamma rays in the mutual annihilation that follows.

Ignoring these trivial cases, we can say the entire Earth is made up of matter—and a good thing, too. If it were made up half of matter and half of antimatter, the one half would instantly annihilate the other and there would be no Earth, only a vast fireball of gamma rays. In fact, it is quite clear that the entire solar system—the entire Galaxy—even the entire local cluster—is matter. Otherwise we would detect far more gamma-ray production than we do.

Can if be that some galactic clusters are matter and some antimatter? Can it be that two universes were formed at the time of the big bang, one of matter and one of antimatter? We don't know. The whereabouts of the antimatter is as yet an unsolved puzzle. If, however, there are both galactic clusters and antigalactic clusters, each retains its integrity because the expanding universe keeps them apart at greater and greater distances.

Can it be, then, that through some fortuitous event or other, an occasional piece of antimatter is ejected from an antigalactic cluster and eventually enters a galactic cluster—or, for that matter, that an occasional piece of matter is ejected from a galactic cluster and eventually enters an antigalactic cluster?

An antistar in our own galaxy might not be recognized as such from its appearance alone, if there were nothing but a good interstellar vacuum in its neighborhood. Even then, though, it would emit occasional gamma rays, as particles of matter in space reacted with the particles of antimatter emitted by the star, and the two groups of particles underwent mutual annihilation.

No such phenomenon has yet been observed, but smaller bodies are both more numerous and more easily ejected than larger ones and there could be in our galaxy occasional objects of planetary or asteroidal size that are antimatter. Might one of them strike the sun without warning? After all, the body might be too small to be seen at a large distance. Even if seen, it might not be possible to recognize it as antimatter until after the strike.

Nevertheless, there isn't much reason to worry about these things. We don't as yet have any evidence that would lead us to suppose that sizable chunks of antimatter are roaming our galaxy. Even if there were, the chances of their striking the sun would probably be no higher than are those of mini-black holes.

Even if a glob of antimatter did strike the sun, the damage it would do is sure to be far more sharply limited than would be the case with a mini-black hole of equal mass. The mini-black hole is permanent and could grow indefinitely at the expense of the sun; the chunk of antimatter, on the other hand, can do no more than annihilate a portion of the sun equal to its own mass and then disappear.

There remains yet a third class of objects that might arrive in the neighborhood of the solar system without being seen much before their arrival. They are neither black holes nor antimatter, but are quite ordinary objects that have escaped our attention simply because they are small.

We can reason out their existence as follows:

I have already said that in any class of astronomical bodies, the small members of the class outnumber the large members. Thus, small stars are more numerous than large ones.

Stars that are roughly the size of the sun (which is a star of intermediate size) make up only about 10 percent of all the stars we see. Giant stars with fifteen times or more the mass of the sun are far fewer. There are a hundred sunlike stars for every such giant. On the other hand, small stars with half the mass of the sun or less make up fully three-quarters of all the stars in the universe, judging from their common occurrence in our near neighborhood.*

* Such small stars are very dim and cannot be seen at great distances. We get a true idea of their frequency, therefore, only by studying our own neighborhood where they are close enough to be seen. At great distances, we see only the large, bright stars and get a false idea of the makeup of the universe.

A body that is only about a fifth the mass of our sun has just barely enough mass to break down the atoms at its center and to start nuclear reactions going. Such a body heats up to a bare red heat and can be seen only faintly, even if fairly close to us as stellar distances go.

Yet there is no reason to think that there is some lower limit in the formation of objects and that this lower limit just happens to coincide with the mass at which nuclear reactions start. There may have been numbers of "substars" that have formed, bodies that are too small to start nuclear reactions at their core, or start them only to the extent of warming up to less than red-heat.

We would recognize such nonshining bodies as planets if they were part of a solar system, and perhaps that is how we ought to view them —as planets that formed independently and owe allegiance to no star, but circle the galactic nucleus independently.

Such "free-planets" may very likely have been formed in far greater numbers than stars themselves and may be very common objects—and yet remain unseen by us, just as the planets of our own solar system would remain unseen, close as they are, did they not happen to reflect light from the nearby sun.

What are the chances, then, of one of these free-planets entering our solar system and creating havoc?

The largest free-planets should be at least as common as the smallest stars, but considering the vastness of interstellar space, this is not common enough for there to be any great chance at all of their encountering us. Smaller free-planets should be more numerous and still smaller ones still more numerous. It follows that the smaller such an object, the greater the chance of its encountering the solar system.

It is quite likely that free-planets of asteroidal size are much more likely to invade the solar system than are either the problematically existing mini-black holes or antimatter. But then, free-planets are far less dangerous than either of the other two objects. Mini-black holes would absorb matter indefinitely should they strike the sun, while antimatter would annihilate matter. Free-planets, made of ordinary matter, would merely evaporate.

If we were to become aware of an asteroid en route to making a close encounter with the sun, we might not be able to tell whether the object is an invader from interstellar space or one of our own home-grown variety that we hadn't happened to notice till then, or that had had its orbit perturbed into a collision course.

It may be that such invading objects have passed through the solar system innumerable times without doing any damage at all. Some small objects of the outer solar system, with suspiciously irregular orbits, may conceivably be free-planets captured en route. These could include Neptune's outer satellite, Nereid; Saturn's outermost satellite, Phoebe; and the curious object, Chiron, discovered in 1977, which orbits the sun in an elliptical orbit lying between those of Saturn and Uranus.

For all we know, in fact, Pluto and its satellite (the latter discovered in 1978) may have been a tiny, independent "solar system" that was captured by the sun. This would make the unusual inclination and eccentricity of Pluto's orbit less surprising.

There remains one other possible type of encounter with objects in interstellar space—encounters with objects so small that they are dust particles or individual atoms. Interstellar clouds of such dust and gas are common in space, and not only can the sun "collide" with such objects, but it undoubtedly has done so on a number of occasions in the past. The effect on the sun of such collisions is negligible to all appearances, but not necessarily so to us. This is a subject to which I will return on a more appropriate occasion later in the book.

Chapter 6

The Death
of the Sun

The Energy-Source

The possible catastrophes of the second class, arising through the invasion of our solar system by objects from without, prove to be of no particular consequence. They are, in some cases, of such low probability in nature, that it is far more likely that we will be first overtaken by a catastrophe of the first class, such as the formation of a new cosmic egg. In other cases, the invasions would seem to be of higher probability, but of lower potentiality for damage to the sun.

Can we then eliminate the reasonable possibility of catastrophes of the second class altogether? Can we decide that our sun is forever safe—or at least is safe while the universe lasts?

Not at all. Even if there is no intrusion from the outside, there is reason to suppose the sun is not safe, and that a catastrophe of the second class, involving the very integrity of the sun, is not only possible but inevitable.

In prescientific times, the sun was widely viewed as a beneficent god, on whose friendly light and warmth humanity, and indeed all life, depended. Its movements in the heavens were closely watched and its path across the sky was seen to rise higher until it reached a peak on June 21 (the summer solstice in the northern hemisphere). It then sank lower in the sky till it reached a trough on December 21 (the winter solstice) and the cycle was then repeated.

Even in prehistoric cultures there seem to have been ways of checking the position of the sun with considerable accuracy; the stones of Stonehenge, for example, seem to be so aligned as to mark out, among other things, the time of the summer solstice.

Naturally, before the true nature of the movements and orientation of the Earth was understood, there could be no confidence that in any one particular year, the sun, as it lowered toward the winter solstice, might not continue to lower indefinitely, disappear, and bring all life to an end. Thus, in the Scandinavian myths, the final end is heralded by the "Fimbulwinter" when the sun disappears and there is a terrible period of darkness and cold that lasts three years —after which is Ragnarok and the end. Even in sunnier climes where faith in the perpetual beneficence of the sun would naturally be stronger, the time of the winter solstice, when the sun ceased its decline, turned, and began to ascend the heavens once more was the occasion of a vast outpouring of relief.

The solstice celebration most familiar to us from ancient times was that of the Romans. The Romans believed that their agricultural god, Saturn, had ruled the land during an early golden age of rich crops and plentiful food. The week of the winter solstice, then, with its promise of a return of summer and of the golden time of Saturnian agriculture, was celebrated with a "Saturnalia" from December 17 to 24. It was a time of unrelieved merriment and joy. Businesses closed so that nothing would interfere with the celebration, and gifts were given all around. It was a time of brotherhood, for servants and slaves were given their temporary freedom and were allowed to join in the celebration with their masters.

The Saturnalia did not disappear. As Christianity gained more and more power in the Roman Empire, it became clear that it could not hope to defeat the joy at the birth of the sun. Some time after A.D. 300, therefore, Christianity absorbed the celebration by arbitrarily declaring December 25 the day on which Jesus was born (something for which there is absolutely no biblical warrant). The celebration of the birth of the sun was thus converted into a celebration of the birth of the Son.

Naturally, Christian thought could not allow godhood to any object in the visible universe, so that the sun was demoted from its divine position. The demotion was minimal, however. The sun was considered a perfect sphere of heavenly light, unchanging and perpetual, from the time God called it forth on the fourth day of Creation until such time, in the uncertain future, as it would please God to bring it to an end. While it existed, it was, in its brilliance and in its unchanging perfection, the most unmistakable visible symbol of God.

The first intrusion of science upon this mythic picture of the sun was Galileo's discovery in 1609 that there are spots on the sun. His observations clearly showed that the spots were part of the solar surface and not clouds obscuring that surface. With the sun no longer perfect, doubts gradually grew as to its perpetuity, too. The more scientists learned about energy on Earth, the more they wondered about the source of the energy of the sun.

In 1854, Helmholtz, one of the important discoverers of the law of conservation of energy, realized it was vital to discover the source of the sun's energy, or the conservation law could not possibly hold. The one source that seemed to him reasonable was the gravitational field. The sun, he suggested, was steadily contracting under the pull of its own gravity, and the energy of that inward-falling motion of all its parts was converted into radiations. If this were so, and if the energy supply of the sun was finite (as it was clear it would have to be), then there had to be both a beginning of the sun and an ending.*

In the beginning, according to Helmholtz's notion, the sun must have been a very thin cloud of gas and its slow contraction under a still not very intense gravitational field would produce little radiant energy. It was only as contraction continued and as the gravitational field, while remaining unchanged in total strength, was concentrated into a smaller volume and therefore grew more intense, that the contraction became rapid enough to deliver the kind of energy with which we are familiar.

It was only about 25 million years ago that the sun contracted to a

* Indeed, if the conservation law holds, *any* source of the sun's energy supply, gravitational or not, must be finite and must come to an end. The law of conservation of energy means, therefore, that the sun must be born and that it must die; in other words, there was a time when the sun was not the familiar object of today, and there will be a time when it will no longer be the familiar object of today. All that can be under dispute is the details of the process.

diameter of 300 million kilometers (186 million miles) and it was only after that that it shrank to a size smaller than the Earth's orbit. It was only at some point less than 25 million years ago that the Earth could have been formed.

In the future, the sun would have to die, for it would eventually contract to the point where it could contract no more and then its source of energy would be consumed and it would no longer radiate, but would cool off and become a cold, dead body—which would certainly be a final catastrophe for us. Considering that it had taken the sun 25 million years to shrink from the size of Earth's orbit to its present size, it might seem surely that it would sink to nothing in about 250,000 years and that that would be all the time left for life on Earth.

Geologists who studied the very slow changes of the Earth's crust were convinced the Earth had to be older than 25 million years. Biologists, who studied the equally slow changes of biological evolution, were also convinced of this. Nevertheless, there seemed no way out of Helmholtz's reasoning but to repeal the law of conservation of energy, or to find a new and larger energy-source for the sun. It was the second alternative that saved the day. A new energy-source was found.

In 1896, the French physicist Antoine Henri Becquerel (1852–1908) discovered radioactivity and it quickly turned out that there was an undreamed-of and enormous energy supply within the nucleus of the atom. If somehow the sun could tap this energy supply, it would not be necessary to suppose it to have been continually shrinking with time. It could radiate at the expense of nuclear energy for extended periods, perhaps, without changing its size much.

Just saying that the sun (and, by extension, the stars generally) are powered by nuclear energy does not, in itself, carry conviction. Precisely how is this nuclear energy made available to the sun?

As long ago as 1862, the Swedish physicist Anders Jonas Angstrom (1814–74) had detected hydrogen in the sun spectroscopically. It gradually came to be known that this simplest of all elements was very common in the sun. By 1929, the American astronomer Henry Norris Russell (1877–1957) showed that, in fact, the sun was predominantly hydrogen. We now know it to be 75 percent hydrogen by mass and 25 percent helium (the second simplest element), with other, more complicated atoms present in only small amounts of fractions of a percent. It is clear from that alone that if there are nuclear reactions taking place in the sun that are responsible for its radiant energy, those reactions must involve hydrogen and helium. Nothing else is present in sufficient quantity to count.

Meanwhile, in the early 1920s, the English astronomer Arthur S. Eddington (1882–1944) demonstrated that the temperature at the center of the sun was in the millions of degrees. At this temperature atoms break down, the electrons on the outskirts are stripped away, and the bare nuclei can slam into each other with such force as to initiate nuclear reactions.

The sun does begin as a thin cloud of dust and gas, as in the Helmholtz hypothesis. It does slowly contract, giving off radiant energy in the process. It is not, however, until it shrinks to something like its present size that it grows hot enough at its core to initiate the nuclear reactions and to begin to shine in its present sense. Once that occurs, it retains its size and its radiant intensity for a long time.

Finally, in 1938, the German-American physicist Hans Albrecht Bethe (1906–), using laboratory data concerning nuclear reactions, showed the probable nature of the reactions taking place in the sun's core to produce its energy. It involved the conversion of hydrogen nuclei into helium nuclei ("hydrogen fusion") by way of a number of well-defined steps.

Hydrogen fusion supplies an adequate amount of energy to keep the sun shining at its present rate for an extended period of time. Astronomers are quite satisfied now that the sun has been shining in its present fashion for nearly 5 billion years. Indeed, it is now thought that the Earth and the sun, and the solar system in general, have been existing in a form recognizable as that in which they exist today for about 4 billion years. This satisfies the needs of geologists and biologists for time in which to allow the changes they have observed to have taken place.

It also means that the sun, the Earth, and the solar system in general can continue to exist (if not interfered with from outside) for billions of additional years.

Red Giants

Even though nuclear energy powers the sun, this merely delays the end. Though the energy supply lasts billions of years rather than millions, it must come to an end eventually.

Until the 1940s, it was assumed that whatever the energy source of the sun, the gradual diminution of that source meant the sun would eventually cool off and that in the end it would dim and darken so that the Earth would freeze in an endless Fimbulwinter. New methods for studying stellar evolution arose, however, and that catastrophe-of-cold proved an inadequate picture of the end.

A star is in balance. Its own gravitational field produces a tendency to contract, while the heat of the nuclear reactions at its core produces a tendency to expand. The two balance each other, and as long as the nuclear reactions continue, an equilibrium is maintained and the star remains visibly unchanged.

The more massive a star, the more intense its gravitational field and the greater its tendency to contract. In order for such a star to remain in volume equilibrium, it has to undergo nuclear reactions at a greater rate in order to develop the higher temperature needed to balance the greater gravity.

The more massive a star, therefore, the hotter it must be and the more rapidly it must consume its basic nuclear fuel, hydrogen. To be sure, a more massive star contains more hydrogen, to begin with, than a less massive star does, but that does not matter. As we consider more and more massive stars, we find that the rate at which the fuel must be expended to balance the gravity goes up considerably faster than the hydrogen content does. That means that a massive star uses up its large hydrogen supply faster than a smaller star uses up its lesser hydrogen supply. The more massive a star the more rapidly it consumes its fuel and the more rapidly it goes through the various stages of its evolution.

Suppose, then, one studies clusters of stars—not globular clusters which contain so many stars that the individual ones cannot be conveniently studied, but "open clusters" containing only a few hundred to a few thousand stars, spread sufficiently far apart to allow for individual study. There are about a thousand of such clusters visible in the telescope and some, like the Pleiades, are close enough so that the brighter members are visible to the naked eye.

All the stars in an open cluster were, presumably, formed more or less at the same time out of a single vast cloud of dust and gas. From that same starting point, however, the more massive ones would have progressed farther on the evolutionary path than the less massive ones, and a whole spectrum of positions on that path could be obtained. The path would, in actual fact, be marked out if temperature and total brightness are plotted against mass. With that as a guide, astronomers can then make use of their increasing knowledge concerning nuclear reactions to understand what must happen inside a star.

As it turns out, although a star must cool off at the end, it goes through a long period during which it actually grows warmer. As hydrogen is converted to helium in the core of a star, the core becomes richer and richer in helium, and therefore becomes more dense. The increasing denseness intensifies the gravitational field in the core which contracts and grows hotter in consequence. The entire star gradually warms for that reason so that while the core contracts, the star as a whole expands slightly. Eventually, the core gets so hot that new nuclear reactions can take place. The helium nuclei within it begin combining to form new and more complex nuclei of the higher elements, such as carbon, oxygen, magnesium, silicon, and so on.

By now, the central core is so hot that the equilibrium is completely overbalanced in the direction of expansion. The star as a whole begins to grow larger at an accelerated pace. As it expands, the total energy radiated by the star increases, but that energy is spread over a vast surface that increases in size even more rapidly. Therefore, the temperature of any individual portion of the rapidly increasing surface goes down. The surface cools to the point where it glows only red-hot instead of white-hot, as in the star's youth.

The result is a "red giant." There are such stars now in the sky. The star Betelgeuse in Orion is one example and Antares in Scorpio is another.

All stars get to the red-giant stage sooner or later; the more massive stars do so sooner, the less massive stars later.

There are some stars that are so huge, massive, and luminous that they will remain in the stable hydrogen-fusing stage (usually called "the main sequence") for less than a million years before swelling into a red giant. There are other stars so small, unmassive and dim, that they will remain on the main sequence for as long as 200 billion years before becoming red giants.

The size of the red giants also depends on mass. The more massive a star, the more voluminously it swells. A really massive star would expand to a diameter many hundreds of times that of the present diameter of our sun, while very small stars would expand to perhaps only a few dozen times its diameter.

Where on this scale is our sun to be found? It is a star of intermediate mass, which means that it has a lifetime on the main sequence that is of intermediate length. It will, eventually, become a red giant of intermediate size. For a star of the sun's mass, the total length of time it will spend on the main sequence, fusing hydrogen quietly and steadily, is perhaps as long as 13 billion years. It has already remained on the main sequence for nearly 5 billion years, which means that the remaining time it has at its disposal is a bit over 8 billion years. During all this time, the sun (as any star would) is undergoing a slow warming. In the last billion years or so of its main sequence, the warming will surely have reached the stage where the Earth will become too hot for life. Consequently, we can look forward to only 7 billion years, at most, during which there will be a life-giving sun worthy of a Saturnalia.

While 7 billion years is not exactly a short period, it is a much shorter period than that required for the coming of a catastrophe of the first class.

At the time the sun begins to climb toward the red-giant stage and life on Earth becomes impossible, there may still remain nearly a trillion years before the coming of the next cosmic egg. It would seem that the entire stay of the sun on the main sequence may be not much more than 1 percent of the life of the universe from cosmic egg to cosmic egg.

By the time, then, that the Earth is no longer a fit abode for life (after having served so for some 10 billion years), the universe as a whole will not be very much more aged than it is now and there will be many generations of stars and planets, yet unborn, waiting to play their role in the cosmic drama.

Assuming that humanity is still in existence on Earth 7 billion years from now (a by-no-means easy assumption, of course), it may well seek to evade this purely local catastrophe and to continue to occupy a still flourishing universe. Evasion won't be easy since there will certainly be no refuge anywhere on Earth. When the sun reaches the peak of its voluminous red gianthood, it will extend to somewhat more than 100 times its present diameter, so that both Mercury and Venus will be engulfed within its substance. Earth may remain outside the swollen bulk of the sun, but, even if the Earth does this, the enormous heat it will receive from the giant sun is quite likely to vaporize it.

Yet even so, all is not lost. There is, at least, ample warning. If humanity survives those billions of years, it will know for all those billions that it will have to plan an escape somehow. As its technological competence increases (and considering how far it has come in the last two hundred years, imagine how far it might go in the course of seven billion) an escape may become possible.

Though the inner solar system will be devastated as the sun expands, the giant planets of the outer solar system, together with their satellites, will suffer less. Indeed, they may, from the human standpoint, experience changes for the better. Humanity may be able to spend considerable time and skill redesigning some of the larger satellites of Jupiter, Saturn, Uranus, and Neptune, in order to make them fit for human habitation. (The process is sometimes called "terra-forming.")

There will be plenty of time to relocate. By the time the sun's expansion begins to speed up and Earth begins to undergo the final bake into irrevocable desert, humanity may be established on a dozen of the outer worlds of the solar system from such satellites of Jupiter as Ganymede and Callisto, out to perhaps Pluto itself. There human beings may be warmed by the large red sun in the sky but not overheated. Indeed, from Pluto, the solar red giant will not look very much larger than the sun does now in Earth's sky.

What's more, it is likely that human beings may establish artificial structures in space that are capable of housing settlements made up of from ten thousand to ten million human beings, each settlement ecologically complete and independent. Nor need these be a product of billions of years of enterprise, since there is every indication that we have the technological capacity to build such settlements now and could fill the sky with them in a matter of a few centuries. Only political, economic, and psychological factors stand in the way (though that's a big "only").

Thus, the catastrophe will be avoided, and humanity, on new worlds, both natural and artificial, can continue to survive.

Temporarily, at any rate.

White Dwarfs

Once hydrogen-fusion is no longer the main source of a star's energy, that star can maintain itself as a large object for only a comparatively short additional period. The energy obtained by fusing helium to larger nuclei and those to still larger ones comes, in total, to not more than 5 percent of what was available from fusing hydrogen. After a comparatively short time, therefore, the ability of the red giant to keep itself distended against the pull of gravity falters. The star begins to collapse.

The lifetime of the red giant and the nature of its collapse depend upon the mass of the star. The larger the mass, the faster the red giant will use up the last dregs of energy available to itself through fusion and the shorter-lived it will be. What's more, the larger the mass, the greater and more intense the gravitational field and, therefore, the more rapid the contraction when it comes.

When a star contracts, there is still considerable hydrogen in its outer layers where nuclear reactions have not been taking place and where the hydrogen has therefore remained untouched. The contraction will heat up the entire star (now it is gravitational energy being converted into heat, à la Helmholtz, not nuclear energy) and so fusion begins in those outer layers. The process of contraction thus coincides with a period of brightening on the outside.

The more massive the star, the more rapid the contraction, the more intense the heating in the outside layers, the more hydrogen there is to fuse and the more rapidly it fuses—and the more violent the results. In other words, a small star would contract quietly, but a large star would undergo enough fusion in its outermost layers to blow off some of its outer mass into space, and do so more or less explosively, leaving only the inner regions to contract.

The more massive the star, the more violent the blowoff. If the star is sufficiently massive, the red-giant stage comes to an end in a violent explosion of unimaginable magnitude, during which a star can briefly glow with a light equal to many billions of times the intensity of an ordinary star; with a glow, in short, equal to an entire galaxy of nonexploding stars. In the course of such an explosion, called a "supernova," up to 95 percent of the matter of a star can be blasted into outer space. What is left over will contract.

What happens to the contracting star that doesn't explode, or to that portion of an exploding star that remains behind and contracts? In the case of a small star that never heats up sufficiently in the course of contraction to explode, it will contract until it is of mere planetary dimensions, while retaining all or almost all its original mass. Its surface is blazing white-hot, considerably hotter than the surface of our sun right now. From a distance, such a contracted star seems dim, however, because the blaze of light comes from such a small surface as not to amount to very much in total. Such a star is a "white dwarf."

Why doesn't the white dwarf continue to shrink? In a white dwarf, the atoms are broken and the electrons, no longer forming shells around central atomic nuclei, form a kind of "electron gas" which can only contract so far. It keeps the matter of the star distended, at least to planetary size, and can do so indefinitely.

Now the white dwarf finally cools, very slowly, and ends its life by becoming too cool to radiate light so that it is then a "black dwarf."

When a star contracts to a white dwarf, it may, if it is not very small, blow away the outermost regions of its red-giant self, in a mild explosion of no great moment as it contracts, losing in this way up to a fifth of its total mass. Seen from a distance, the white dwarf that forms would seem to be surrounded by a luminous fog, almost like a smoke ring. Such an object is called a "planetary nebula" and there are a number of these in the sky. Gradually, the cloud of gas drifts outward in all directions, becomes dimmer, and vanishes into the general thin matter of interplanetary space.

When a star is massive enough to explode violently in the process of contraction, the remnant that does contract may still be too massive—even after the loss of considerable mass in the explosions—to form a white dwarf. The more massive the contracting remnant, the more tightly squeezed in upon itself is the electron gas, and the smaller the white dwarf.

Finally, if there is enough mass, the electron gas cannot withstand the pressure upon itself. The electrons are squeezed into the protons present in the nuclei that are wandering about in the electron gas, and neutrons are formed. These are added to the neutrons that already exist in the nuclei and the star then consists primarily of neutrons and nothing else. The star contracts until those neutrons are in contact. The result is a "neutron star" which is only the size of an asteroid, perhaps ten or twenty kilometers across, but which preserves the mass of a full-sized star.

If the contracting remnant of the star is still more massive, then not even the neutrons will be able to withstand the gravitational inpull. They will smash and the remnant will contract further into a black hole.

What, then, will be the fate of the sun after it reaches the red-giant stage?

It may remain a red giant for a couple of hundred million years—a very brief interval on the scale of stellar lifetimes but allowing an extended period for civilization to develop on the terra-formed outer worlds and in the space settlements—but then the sun will contract. It will not be large enough to explode violently so that there will be no danger that in a day or week of fury the solar system will be cleansed of life out to the orbit of Pluto and beyond. Not at all. The sun will simply contract, leaving behind at most a thin film of its outermost layer, making of itself a planetary nebula.

The cloud of matter will drift by the distant planets that we have imagined to be housing the descendants of humanity in those far-future times and will probably not offer much of a danger to them. It will be a very thin gas, even to begin with, and if, as may well be true, the human colonies live underground or within domed cities, there may be no adverse effect whatever.

The real problem will be the shrinking sun. Once the sun has shrunk to a white dwarf (it is not massive enough to form a neutron star and certainly not a black hole) it will be no more than a tiny dot of light in the sky. Seen from the satellites of Jupiter, if human beings have managed to establish themselves that close to the sun during its red-giant stage, it will be only 1/4000 as bright as the sun appears to us on Earth now, and it will deliver only that fraction of energy, too.

If the human settlements in the outer solar system depend upon the sun for energy, they will not be able to get enough energy to maintain their societies once the sun has become a white dwarf. They will have to move in considerably closer, and they won't be able to do that if they require a planet for the purpose, since the planetary bodies of the inner solar system will have been ruined or destroyed outright in the preceding red-giant phase of the sun's existence. That will leave only the artificial space settlements to serve as a refuge for humanity in the time to come.

When such settlements are first built (perhaps in the next century or so), they will move in orbits about the Earth, using solar radiation as their energy source and the moon as their source of most of their raw materials. Some essential light elements—carbon, nitrogen, and hydrogen—which are not present in appreciable quantities on the moon will have to be obtained from Earth.

Eventually, it is already foreseen, such space settlements will be built in the asteroid belt where it will be easier to get those vital lighter elements, without having to indulge in a dangerous dependence upon Earth.

It may be that as space settlements become more self-contained and more mobile, and as humanity foresees more clearly the difficulty of remaining tied to planetary surfaces in view of the viscissitudes that will overtake the sun in its latter days, the space settlements may become the preferred abode of humanity. It is quite conceivable that long before there is any question of the sun giving us any trouble, most or all of humanity will be utterly free of the surfaces of the natural planets and will live in space—in worlds and in environments of their own choosing.

There may then be no question of terra-forming outer worlds in order to survive the red-gianthood of the sun. That might, by then, seem a clumsy solution for which there would be no necessity. Instead, as the sun grows very gradually hotter, the space settlements will adjust their orbits accordingly and drift very slowly farther out.

This is not a difficult thing to imagine. The orbit of a world like the Earth is almost impossible to change, because it has so huge a mass and therefore so great a momentum and an angular momentum that adding or subtracting a sufficient sum to alter the orbit significantly is an impractical undertaking. And the mass of Earth is necessary, if it is to have enough of a gravitational field to hold an ocean and an atmosphere to its surface and thus make life possible.

In a space settlement, the total mass is insignificant compared to the Earth since gravitation is not used to retain water, air, and everything else. Instead, it is all retained by being mechanically closed in by an outer wall, and the effect of gravity on the inner surface of that wall will be produced by the centrifugal effect that originates through rotation.

The space settlement, then, can have its orbit changed by the expenditure of a reasonable amount of energy and it can be moved farther from the sun as the sun grows warmer and expands. It can, in theory, move closer to the sun as the sun contracts and supplies less total energy. The contraction, however, will be much more rapid than the previous expansion. What's more, for all the space settlements that may exist in the red-giant stage of the sun to move into the neighborhood of the white dwarf will perhaps constrict them into a smaller volume than they care for. They may have become accustomed, for billions of years, to the unlimited spaces of a large solar system.

But then it is not beyond the bounds of conceivability that long before the time of white-dwarfhood comes, the space settlers will have developed some form of hydrogen-fusion power stations as a source of energy and that they would then be independent of the sun. They might, in that case, choose to leave the solar system altogether.

If a significant number of space settlements leave the solar system, becoming self-propelled "free-planets," it will mean that humanity would be free of the danger of catastrophes of the second class and might continue to live on (and to spread through the universe to an indefinite degree) until the coming of the universal contraction into a cosmic egg.

Supernovas

The chief reasons why the death of the sun (a death in the sense that it will become completely different from the sun we know) need not be a catastrophe for the human species are (1) that the inevitable expansion and subsequent contraction of the sun will come so far in the future that by then human beings will surely have developed the technological means to escape, assuming they are still surviving; and (2) that the changes are so predictable that there is no chance of being caught by surprise.

What we must consider now, then, are possible ways in which catastrophes of the second class (involving the sun or, by extension, a star) might catch us by surprise and, worse yet, do so in the near future before we have the chance to develop the necessary technological defenses.

There are stars that undergo catastrophic changes, for instance; that brighten in the process even from invisibility and then dim again, sometimes even to invisibility. These are the "novas" (from the Latin word for "new" since they seemed to be new stars to ancient astronomers who lacked telescopes). The first of these was mentioned by the Greek astronomer Hipparchus (190–120 B.C.).

Unusually bright novas are the "supernovas" we have already referred to, a name first used by the Swiss-American astronomer, Fritz Zwicky (1898–1974). The first one to be discussed in detail by European astronomers was the supernova of 1572.

Suppose, for instance, that it is not the sun that approaches the end of its life on the main sequence, but some other star. Although our sun is still in early middle age, some nearby star might be old and on the point of death. Might a nearby supernova blaze out suddenly, catch us by surprise, and affect us catastrophically?

Supernovas are not common; only one star in a hundred is capable of exploding as a supernova and of them only a few are in the final stages of their lifetime and of them fewer still are close enough to be seen as unusually bright stars. (Before the invention of the telescope, it took an unusually bright star to obtrude itself on the notice of observers as something that had appeared where no star had been visible before.) Still supernovas can appear and in the past have done so—without warning, of course.

One remarkable supernova to appear in the sky in historic times showed up on July 4, 1054—undoubtedly the most tremendous bit of fireworks known to celebrate the Glorious Fourth, albeit 722 years before the event. This supernova of 1054 was observed by Chinese astronomers, but *not* by European and Arabic astronomers.*

The supernova appeared as a new star, blazing out in the constellation Taurus with a fury that caused it to exceed Venus in brightness. Nothing in the sky was brighter than the new star, except for the sun and the moon. It was so bright it could be seen by daylight—and not just for a brief period but for day after day over a period of three weeks. Slowly then it began to fade; but it was nearly two years before it was too faint to be seen by the naked eye.

In the spot where the ancient Chinese astronomers reported this extraordinary apparition, there is now a turbulent cloud of gas called the "Crab nebula," which is about 13 light-years in diameter. The Swedish astronomer Knut Lundmark first suggested in 1921 that this might be a surviving remnant of the supernova of 1054. The gases of the Crab nebula are still moving outward at a speed that, calculated backward, shows that the explosion driving them took place just at about the time the new star appeared.

Bright as that supernova was in the sky of 1054, it delivered to the Earth not more than a hundred-millionth of the light of the sun, and that is scarcely enough to affect human beings in any way, especially since it only remained at that level for a few weeks.

It is not, however, just the total light that counts but the distribution. Our sun delivers some very active radiation in the form of X-rays, but a supernova has a much larger percentage of its radiant energy in the X-ray region. The same is true of cosmic rays, another form of high-energy radiation we will return to later.

In short, though the light of the supernova of 1054 was so dim compared to the sun, it may have rivaled the sun in its output of Earth-striking X-rays and cosmic rays, at least in the initial weeks of the explosion.

Even so, that was not dangerous. Although, as we shall see, the influx of energetic radiation can have a deleterious effect on life, our

* Astronomy in Europe was at a low ebb at that time and those who did watch the heavens may have been too firmly convinced of the ancient Greek doctrine of the unchangeability of the heavens to accept the evidence of their eyes.

atmosphere protects us from unreasonable quantities of it, and neither the supernova of 1054 nor the sun itself is unduly dangerous to us under our blanket of protective air. Nor is this merely speculation. The fact is that Earth's load of life went right through that critical year of 1054 with no detectable ill-effects.

Of course, the Crab nebula is not very near to us. It is about 6,500 light-years away.* A still brighter supernova appeared in the year 1006. From the reports of Chinese observers, it would seem to have been possibly as much as a hundred times as bright as Venus, and a respectable fraction of the brightness of the full moon. There are references to it even in a couple of European chronicles. It was only 4,000 light-years away.

Since 1054, there have been only two visible supernovas in our sky. A supernova occurred in Cassiopeia in 1572 that was almost as bright as the one of 1054, but was farther away in space. Finally, there was a supernova in Serpens in 1604, that was considerably less bright than any of the other three I've mentioned, but also considerably farther away.†

Some supernovas could have taken place in our galaxy since 1604 and have remained invisible, hidden behind the vast clouds of dust and gas that clog the outskirts of the galaxy. We can, however, detect the remants of supernovas in the form of rings of dust and gas, like that of the Crab nebula, but usually thinner and wider, which give a hint of supernovas that have exploded without being seen, either because they were hidden or because they occurred too far back in time.

A few wisps of gas marked by microwave emission and called Cassiopeia A seem to mark a supernova that exploded in the late 1600s. If so, that is the most recent supernova known to have exploded in our galaxy though it could not be seen at the time. This explosion may have been considerably more spectacular than the supernova of 1054 if viewed at the same distance, judging from the radiation given off now by its remnants. It was, however, 10,000 light-years away, so that it probably wouldn't have been much brighter than the earlier one—if it could have been seen.

---

* Imagine the fury of an explosion that could create a light brighter than that of Venus from a distance that enormous.
† It is rather frustrating to astronomers that while there were two supernovas visible to the naked eye in the space of 32 years just before the invention of the telescope, there has not been one since. Not one! The brightest supernova seen since 1604 was one in 1885, located in the Andromeda galaxy. It grew almost bright enough to be seen by the naked eye, even at the vast distance of that galaxy—but not quite.

A more spectacular supernova than any seen in historic times blazed out in the sky perhaps 11,000 years ago, at a time when, in some parts of the world, human beings were soon to develop agriculture. What is left of that supernova now is a shell of gas in the constellation of Vela, first detected in 1939 by the Russian-American astronomer Otto Struve (1897–1963). This shell is called the Gum nebula (named for the Australian astronomer Colin S. Gum who first studied it in detail in the 1950s).

The center of the shell is only 1,500 light-years from us, which makes it, of all the known supernovas, the one that exploded nearest us. One edge of the still-expanding-and-thinning shell of gas is only about 300 light-years from us now. It may reach us in about 4,000 years or so, but it will be such thinly spread-out matter that it should not affect us in any significant way.

When that nearby supernova blasted, it may at its peak have been as bright as the full moon for some days, and we may envy those prehistoric human beings who witnessed that magnificent sight. Nor did that, either, seem to harm life on Earth.

Yet even the Vela supernova was 1,500 light-years away. There are stars at less than a hundredth that distance. What if a star really close to us unexpectedly went supernova? Suppose one of the Alpha Centauri stars, only 4.4 light-years away, went supernova—what then? If a bright supernova flashed into existence 4.4 light-years away, as bright as a supernova ever gets, it would blaze with nearly 1/6 the light and heat of the sun and, for a few weeks, there would be a heat wave such as Earth has never seen.*

Suppose the supernova blazed out at Christmas time as the brightest Star of Bethlehem ever. At that time of year, it would be the summer solstice in the southern hemisphere and Antarctica would be entirely exposed to continuous sunlight. The sunlight would be weak, to be sure, for from Antarctica the sun is close to the horizon even at the solstice. The Alpha Centauri supernova would, however, be high in the sky and would add its quite substantial heat to that of the sun. The Antarctica ice cap would be bound to suffer. The amount of melting would be unprecedented and the sea level would rise measurably, with disastrous effect in many places in the world. Nor would the sea level recede quickly after the supernova had cooled down. It would take years for equilibrium to be restored.

* In the United States and Europe the supernova would be invisible, for Alpha Centauri is a far southern star not visible in northern latitudes, but the hot winds from the south would let us know that *something* had happened.

In addition, Earth would be bathed in X-rays and cosmic rays at intensities it has perhaps never before received, and, after a few years, a cloud of dust and gas, thicker than any it has ever encountered, would envelop it. We will discuss later what effects these events might have but they would surely be disastrous.

The saving grace is that it won't happen. Indeed, it *can't* happen. The brighter of the stars of the Alpha Centauri binary is just about exactly the mass of the sun, and it can no more blow up as a giant supernova, or as any kind of supernova, than our sun can. The most that Alpha Centauri can do is to go red giant, pop off some of its outermost layers as a planetary nebula, and then shrink to a white dwarf.

We don't know when that will happen, for we don't know how old it is, but it can't happen until after it turns red giant, and even if that were to begin to happen tomorrow it would probably remain in the red-giant stage for a couple of hundred million years.

What, then, is the smallest distance at which we could possibly find a supernova?

To begin with, we must look for a massive star; one that is 1.4 times as massive as the sun as an absolute minimum, and one that is considerably more massive than that, if we want a really big show. These massive stars are not common and that is the chief reason that supernovas are no more common than they are. (It is estimated that in a galaxy the size of our own there may be one supernova somewhere in it every 150 years on the average and, of course, few of those are likely to be even moderately close to us.)

The nearest massive star is Sirius, which is 2.1 times the mass of our sun and is 8.63 light-years away, just about twice the distance of Alpha Centauri. Even with that mass, Sirius is not capable of producing a really spectacular supernova. It will someday explode, yes, but it will be a handgun rather than a cannon. Besides, Sirius is on the main sequence. Because of its mass, its total lifetime on the main sequence is only some 500 million years and some of that has clearly been expended. What is left, plus the red-giant stage, must mean, however, that again an explosion is some hundreds of millions of years off.

What we must ask, then, is which is the nearest massive star that is already in the red-giant stage.

The nearest red giant is Scheat in the constellation of Pegasus. It is only about 160 light-years away and its diameter is about 110 times that of the sun. We don't know its mass, but if this is as wide as it is going to get, its mass is very little more than that of the sun and it will not pass into the supernova stage. If, on the other hand, it is more massive than the sun and is still expanding, its supernova stage is yet a long time off.

The nearest really large red giant is Mira, in the constellation of Cetus. Its diameter is 420 times that of the sun, so that if it were imagined to be in place of the sun, its surface would be located in the farther reaches of the asteroid belt. It must be considerably more massive than the sun, and it is about 230 light-years away.

There are three red giants that are larger still and are not very much farther away. These are Betelgeuse in Orion; Antares in Scorpio; and Ras Algethi in Hercules. Each of these is about 500 light-years away.

Of these Ras Algethi has a diameter 500 times that of the sun and Antares one that is 640 times that of the sun. If Antares were imagined in place of the sun, with its center located at the sun's center, its surface would extend beyond the orbit of Jupiter.

Betelgeuse has no fixed diameter because it seems to pulsate. When it is at its smallest it is no larger than Ras Algethi, but it can expand to a maximum of 750 times the diameter of the sun. If Betelgeuse were imagined in place of the sun, its surface would, at maximum, reach out to the midway point between Jupiter and Saturn.

It is probable that Betelgeuse is the most massive of these nearby red giants and its pulsation may be an indication of instability. In that case it may be that of all the stars reasonably near to us, it is closest to supernova and collapse.

Another indication of this is the fact that photographs of Betelgeuse, taken in 1978 in the range of infrared light (light with longer waves than those of red light and therefore not capable of affecting the retina of the eye), show the star to be surrounded by an enormous shell of gas some 400 times the diameter of Pluto's orbit about our sun. It may be that Betelgeuse is already beginning to blow off matter in the first stage of supernovahood.

Without knowing its mass, we can't predict how bright the Betelgeuse supernova would be but it should be of respectable size. What it may lack in intrinsic brightness it would make up for by being at only one-third the distance of the Vela supernova. It may therefore, when it comes, be brighter than the supernova of 1006 and perhaps even rival the Vela supernova. The skies might light up with a new kind of moonlight and the Earth might be bombarded with a greater concentration of hard radiation than it has experienced since the Vela supernova 11,000 years ago.

Since Homo sapiens—and life generally—seems to have survived the Vela supernova handily, there is every hope that it would survive the Betelgeuse supernova as well.*

We cannot, as yet, tell the exact time when Betelgeuse might reach the explosion point. It may be that its present variable diameter is an indication that it is on the point of collapse and that each time it begins, the rising temperature that accompanies the collapse allows a recovery. Eventually, we can suppose, one collapse will go so far that it will set off the explosion. That "eventually" may not be for centuries; on the other hand, it may be tomorrow. In point of fact, Betelgeuse may have exploded five centuries ago and the wave of radiation, traveling toward us all that time, may reach us tomorrow.

Even if a Betelgeuse supernova is the worst we can expect in the reasonably close future, and if we can convince ourselves that it will present us with a fascinating show but with no serious danger, we are still not home free as far as stellar explosions are concerned. The more distant future may hold greater dangers well before the time of the death of our own sun arrives.

After all, the situation of today is not permanent. Every star, including our own sun, is moving. Our sun is constantly moving into new neighborhoods, and the neighborhoods are themselves constantly changing.

With time the various changes may just possibly bring our sun into the near neighborhood of a giant star that will happen to explode into a supernova as it passes us. The fact that the Betelgeuse supernova is the worst we can expect right now is no indication of eternal safety; it is an accident of the moment.

Such a neighboring-star catastrophe is not likely to happen for a long time to come, however. As I have pointed out, stars move very slowly in comparison to the vast distances between them, and it will be a long time before stars now distant from us come significantly closer.

The American astronomer Carl Sagan (1935–) calculates that a supernova may explode within 100 light-years of us at average intervals of 750 million years. If this is so, such nearby explosions may have taken place perhaps six times in the history of the solar system so far and may take place nine times more before the sun leaves the main sequence.

Such an event cannot, however, catch us by surprise. It is not

---

* There is a combination of circumstances, as we shall see later, that may make the situation worse for us.

difficult to tell which stars are approaching. We can tell a red-giant star even at a distance considerably in excess of 100 light-years. It is very likely we will know that there is a chance of such an explosion with an advance warning period of at least a million years and will be able to plan action to minimize or evade the effects of the explosion.

Sunspots

The next question is this: Can we entirely rely on our own sun? Could something go wrong with the sun while it is yet on the main sequence? Might something go wrong in the near future and without warning so that we would lack defenses, or time to deploy them if we had them?

Unless there is something terribly wrong with our present beliefs concerning stellar evolution, nothing very much can go wrong with the sun. As it is now, so it has been for a very long time, and so it will remain for a very long time. Any change in its behavior will have to be so small as to be inconsequential on the solar scale.

But could not variations that are inconsequential on the solar scale be disastrous on the earthly scale? Clearly, yes. A small hiccup in the sun's behavior may be nothing to it and might be unnoticeable if the sun were viewed from the distance of even the nearer stars. The effect on Earth of such a small change, however, may be enough to alter its properties drastically and, if the abnormal spasm were to endure long enough, it might visit us with true catastrophe.

Life as we know it is, after all, a rather fragile thing on the cosmic scale. It does not take a very great temperature change to boil the oceans or freeze them, and in either case make life impossible. Relatively small changes in the solar output would suffice to produce either extreme. It follows, then, that for life to continue the sun must shine with only tiny variations, at most, from its general state.

Since the history of life is a continuous one over more than three billion years as nearly as we can tell, we have the heartening assurance that the sun is a reliable star indeed. Still, the sun might be steady enough to allow life to exist in general, and yet be unsteady enough to put it through some mighty terrible hardships. There have indeed been times in life's history when there seem to have been biological catastrophes and we can't be sure that the sun wasn't responsible. This we will consider later.

If we confine ourselves to historic times, the sun has seemed perfectly stable, at least to casual observers and to astronomers less well-endowed with instruments than those of our sophisticated present day. Are we living in a fool's paradise to suppose this will continue?

One way of telling is to observe other stars. If all other stars are perfectly constant in brightness, then why should we not assume that our sun will also be so, never giving us either too much radiation or too little?

As a matter of fact, though, a few stars visible to the naked eye are *not* steadily bright, but vary, being dimmer at some times and brighter at others. One such star is Algol in the constellation Perseus. No astronomer of ancient or medieval times seems to have referred to its variability, perhaps because of the strength of the Greek belief that the heavens were unchangeable. There is indirect evidence, though, that astronomers may have been aware of the variability even if they didn't like to talk about it. Perseus, in the constellation, was usually pictured as holding the head of the slain Medusa, the demon-monster whose hair consisted of living snakes and whose fatal glance turned men to stone. Algol was pictured as marking that head and it was sometimes called the "Demon star" in consequence. In fact "Algol" itself is a distortion of the Arabic "al ghul" meaning "the ghoul."

One is tempted to suppose that the Greeks were too disturbed by Algol's variability to refer to it openly but exorcised it by making it a demon. The fact of its variability was first noted explicitly in 1669 by the Italian astronomer Geminiano Montanari (1632–87). In 1782, an eighteen-year-old deaf-mute, the Dutch-English astronomer John Goodricke (1764–86), showed that the variability of Algol was absolutely regular, and suggested that it was not truly variable. Instead, he suggested, it had a dim companion star which circled it and, periodically, partially eclipsed it. As it turned out, he was perfectly right.

Earlier, though, in 1596, the German astronomer David Fabricius (1564–1617) had noted a variable star that was much more remarkable than Algol turned out to be. It was Mira, the star I mentioned earlier as a nearby red giant. "Mira" is from a Latin word meaning "cause for wonder" and so it was, in that it varies in brightness to a much greater extent than Algol does, growing so dim at times as to be invisible to the naked eye. Mira also has a much longer and much more irregular period of variation than Algol has. (Again one feels sure that this must have been noted before, but may have been deliberately ignored as too disturbing to accept.)

We can ignore stars like Algol, which undergo eclipses and only *seem* to vary in light. Their case does not indicate any sign of disastrous variability in a star like the sun. We can also ignore the supernovas which occur only in the convulsions of a star undergoing its final collapse, and the ordinary novas, which are white-dwarf stars that have already undergone collapse and are absorbing an unusual quantity of matter from a normal companion star.

That leaves stars like Mira or Betelgeuse, which are "intrinsic variable stars"; that is, stars that vary in the light they emit because of cyclic changes in their structure. They pulsate, in some cases regularly and in others irregularly, growing cooler but larger in the expanding portion of their cycle, and hotter but smaller in the contracting portion.

If the sun were such an intrinsic variable star, life on Earth would be impossible, for the difference in radiation emitted by the sun at different times in its cycle would periodically wash the Earth with unbearable heat and subject it to unbearable cold. We might argue that human beings could protect themselves from such temperature extremes, but it seems unlikely that life would have developed under such conditions in the first place, or that it would have evolved to the period where any species was sufficiently advanced technologically to deal with such variations. Of course, the sun is *not* such a variable star, but might it become one, and might we suddenly find ourselves living on a world with temperature extremes that made it an unbearable horror?

That, fortunately, is not at all likely. In the first place, intrinsic variable stars are not common. There are perhaps only 14,000 known altogether. Even admitting that many such stars go unnoticed because they are too distant to be seen or because they are hidden behind dust clouds, the fact remains that they represent a very small percentage of all stars. The vast majority of stars seem to be as stable and unvarying as the ancient Greeks thought they were.

Furthermore, some intrinsically variable stars are large, bright stars near the end of their stay on the main sequence. Others, like Mira and Betelgeuse, have already left the main sequence and seem to be near the end of their lives as red-giant stars. It is quite likely that pulsation marks the kind of instability that indicates an end to a certain stage of a star's lifetime and the approaching shift to some other stage.

Since the sun is still but a middle-aged star with billions of years to go before the present stage comes to an end, there seems no chance that it will become a variable star for a long time in the future. Even so, there are degrees of variability, and the sun might be, or become, variable to a very tiny degree and yet cause us trouble.

What about sunspots, for instance? Might their presence in varying amounts from time to time indicate a certain small variability in the sun's output of radiation? The spots are known to be distinctly cooler than the unspotted portions of the sun's surface. Might not a spotty sun therefore be cooler than a spotless one and might we not experience the effects here on Earth?

This question grew more important with the work of a German pharmacist, Heinrich Samuel Schwabe (1789–1875), whose hobby was astronomy. He could devote himself to his telescope only in the daylight hours, so he took to observing the neighborhood of the sun to try to detect an unknown planet some thought might be orbiting the sun inside Mercury's orbit. If this were true, it might very well cross the sun's disc periodically and for this Schwabe watched.

He began his search in 1825, and in scanning the Sun's disc, he could not help but note the sunspots. After a while he forgot about the planet and started sketching the sunspots. For seventeen years he did this on every sunny day. By 1843, he was able to announce that the sunspots waxed and waned in number in a ten-year cycle.

In 1908, the American astronomer George Ellery Hale (1868–1938) was able to detect strong magnetic fields inside sunspots. The direction of the magnetic field is uniform through a particular cycle and then reverses in the next one. Taking magnetic fields into account, the time from one sunspot maximum with the field in one direction, to the next maximum with the field in that same direction is twenty-one years.

Apparently, the sun's magnetic field strengthens and declines for some reason and the sunspots are associated with this change. So are other effects. There are "solar flares," sudden temporary brightenings of the surface of the sun here and there, that seem associated with local strengthening of the magnetic field. These grow more common as the sunspots increase in number, since both reflect the magnetic field. Consequently at sunspot maximum we speak of an "active sun" and at sunspot minimum of a "quiet sun." *

Then, too, the sun is always giving off streams of atomic nuclei

---

* The heat of the flares may more than make up for the coolness of the spots, so that a spotted sun may be warmer than an unspotted one.

(chiefly hydrogen nuclei, which are simple protons) and these move outward from the sun at great speeds in all directions. This was dubbed the "solar wind" in 1958 by the American astronomer Eugene Norman Parker (1927–).

The solar wind reaches and passes the Earth and interacts with the upper atmosphere to produce a variety of effects, such as the aurora borealis (or "northern lights"). Solar flares spew out enormous quantities of protons and temporarily strengthen the solar wind. In this way, the Earth is affected much more strongly by rises and falls in solar activity than by any simple change in temperature associated with the sunspot cycle.

The sunspot cycle, whatever its effects on Earth, clearly does not interfere with life in any obvious way. The question, though, is whether the sunspot cycle can ever get out of hand and whether the sun might start see-sawing so violently, so to speak, as to produce a catastrophe. We might argue that since it has never done so in our past, as far as we know, it should not do so in the future. Our confidence in this argument would be stronger if the sunspot cycle were perfectly regular, but it is not. The time between sunspot maxima, for instance, has been recorded to be as short as seven years, or as long as seventeen.

Then, too, the intensity of the maxima is not fixed. The extent of the spottedness of the sun is measured by the "Zürich sunspot number." This counts 1 for each individual sunspot and 10 for each group of sunspots, and multiplies the whole by a figure that varies with the instruments used and with the conditions of observation. If the Zürich sunspot number is measured from year to year it turns out that there have been sunspot maxima with numbers as low as 50, as in the early 1700s and the early 1800s. On the other hand, the maximum in 1959 reached an all-time high of 200.

Naturally, sunspot numbers have been recorded with careful assiduity only since Schwabe's report in 1843, so that the figures we use for the years before that, back to 1700, are perhaps not entirely reliable, and reports from the first century after Galileo's discovery have usually been discounted altogether as too fragmentary.

In 1893, however, the British astronomer Edward Walter Maunder (1851–1928), searching through old records, was astonished to find that what observations of the sun's surface were made between 1645 and 1715 simply did not speak of sunspots. The total number of spots reported for that seventy-year period was less than that reported in any one year now. The finding was ignored at the time, since it seemed easy to suppose that the seventeenth-century data were too fragmentary and unsophisticated to be meaningful, but recent research has borne Maunder out, and the period from 1645 to 1715 is now called the "Maunder minimum."

Not only were the sunspots almost absent in that period, but reports of auroras (which are most common at sunspot maximum when flares burst out all over the sun) almost ceased in that period. What's more, the shape of the corona during total eclipses of the sun, judging from descriptions and drawings in that period, was characteristic of its appearance at sunspot minimum.

Indirectly, the variations in the magnetic field of the sun, as evident in the sunspot cycle, affect the quantity of carbon-14 (a radioactive form of carbon) in the atmosphere. The carbon-14 is formed by cosmic rays striking the Earth's atmosphere. When the sun's magnetic field is expanded during sunspot maximum, it helps protect the Earth against the cosmic ray influx. At sunspot minimum, the magnetic field shrinks and the cosmic rays are not deflected. It follows that carbon-14 in the atmosphere is high at sunspot minimum and low at sunspot maximum.

Carbon (including carbon-14) is absorbed by plant life in the form of the carbon dioxide in the atmosphere. Carbon (including carbon-14) is incorporated into the molecules making up the wood of trees. Fortunately, carbon-14 can be detected and its quantity determined with great delicacy. If very old trees are analyzed, the carbon-14 in each annual ring can be determined, and one can tell from year to year how the carbon-14 varies. It is high at sunspot minimum and low at sunspot maximum and it turns out to have been high all during the Maunder minimum.

Other extended periods of solar inactivity have been found in this way, some lasting for as little as fifty years and some for as long as several centuries. About twelve of them have been detected in historic times since 3000 B.C.

In short, there seems to be a larger sunspot cycle. There are extended minima of very little activity, interspersed by extended periods of oscillations between low and high activity. We happen to have been in one of the latter periods ever since 1715.

What effects does this larger sunspot cycle have on the Earth? Apparently, the dozen Maunder minima that have taken place in historic times do not seem to have interfered catastrophically with human existence. On that basis, it would seem we need not fear recurrence of such extended minima. On the other hand, it does show we don't know as much about the sun as we thought we did. We do not thoroughly understand what causes the ten-year sunspot cycle that now exists and we certainly don't understand what causes the Maunder minima. As long as we don't understand such things, can we be sure that the sun might not at some time go out of control without warning?

Neutrinos

It might help, of course, if we knew what went on inside the sun not just as a matter of theory, but as a matter of direct observation. This might seem to be a useless hope but, as it happens, it isn't quite.

In the early decades of the twentieth century, it became clear that when radioactive nuclei broke down they would often emit speeding electrons. These electrons possessed a wide range of energies that almost never came up to the total amount of energy the nucleus had lost. This seemed to go against the law of conservation of energy.

In 1931, the Austrian physicist Wolfgang Pauli (1900–58), in order to avoid breaking that law, as well as several other conservation laws, suggested that a second particle was always emitted along with the electron and that it was the second particle that contained the missing energy. To account for all the facts of the case, the second particle had to be carrying no electric charge and probably had to be without mass. Without charge or mass, it would be extremely difficult to detect. The Italian physicist Enrico Fermi (1901–54) called it a "neutrino," Italian for "little neutral one."

Neutrinos, assuming they have the properties they were thought to possess, would not readily interact with matter. They would pass through the entire Earth just about as easily as they would pass through the same thickness of vacuum. In fact, they would pass through billions of earths lined up side by side with very little trouble. Nevertheless, every once in a long while, a neutrino could strike a particle under conditions where an interaction would take place. If one were to work with many trillions of neutrinos, all streaming through a small body of matter, a few interactions might take place and these might be detected.

In 1953, two American physicists, Clyde L. Cowan, Jr. (1919–) and Frederick Reines (1918–), worked with the antineutrinos* being

---

* These are like neutrinos but are opposite in certain properties. As a matter of fact, it is an antineutrino and not a neutrino that is given off along with an electron when certain nuclei break down.

given off by uranium-fission reactors. These were allowed to pass through large tanks of water and certain predicted interactions did take place. After twenty-two years of merely theoretical existence, the antineutrino, and therefore the neutrino as well, were shown to exist experimentally.

Astronomical theories concerning the nuclear fusion of hydrogen into helium in the sun's core—the source of the sun's energy—require that neutrinos (*not* antineutrinos) be given off in great quantities, quantities that amount to 3 percent of the total radiation. The other 97 percent is made up of photons, which are the units of radiant energy such as light and X-rays.

The photons make their way to the surface and are finally radiated into space, but this takes a long time, since photons readily interact with matter. A photon that is produced at the sun's core is absorbed very quickly, reemitted, absorbed again, and so on. It could take a million years for a photon to make its way from the core of the sun to its surface, even though it travels at the speed of light between absorptions. Once the photon reaches the surface it has had such a complicated history of absorptions and emissions that it is impossible to tell from its nature what went on in the core.

It is quite different where the neutrinos are concerned. They travel at the speed of light, too, since they are massless. However, because they so rarely interact with matter, neutrinos produced at the sun's core pass right through the sun's matter, reaching its surface in 2.3 seconds (and losing only 1 in 100 billion through absorption in the process). They then cross the vacuum of space and in 500 more seconds reach the Earth if they happen to be aimed in the right direction.

If we could detect these solar neutrinos here on Earth, we would have some direct information concerning events in the sun's core some eight minutes earlier. The difficulty lies in detecting the neutrinos. This task has been undertaken by the American physicist Raymond Davis, Jr., who took advantage of the fact that sometimes a neutrino will interact with a variety of chlorine atoms to produce a radioactive atom of the gas, argon. The argon can be collected and detected even if only a few atoms are formed.*

For the purpose, Davis made use of a huge tank containing 378,000 liters (100,000 gallons) of tetrachloroethylene, a common cleaning fluid that happens to be rich in chlorine atoms. He placed it deep in

* This possibility was first pointed out in the late 1940s by the Italian-Canadian physicist Bruno M. Pontecorvo (1913–).

the Homestake gold mine in Lead, South Dakota, where there were 1.5 kilometers (1 mile) of rock between the tank and the surface. All that rock would absorb any particles coming from space except neutrinos.

It was then only a matter of waiting for argon atoms to form. If accepted theories of events in the sun's core were correct, then a certain number of neutrinos should be formed each second, of which a certain percentage should reach the Earth; of these, a certain percentage should pass through the tank of cleaning fluid, and among the latter, a certain percentage should interact with chlorine atoms to form a certain number of argon atoms. From fluctuations in the rate at which argon atoms were formed, and from other properties and variations of the interaction generally, conclusions might be drawn concerning the events at the sun's core.

Almost at once, however, Davis had cause for astonishment. Very few neutrinos were detected; far fewer than had been expected. At the most, only one-sixth as many argon atoms were formed as should have been formed.

Clearly astronomic theories as to events at the sun's core seem to require revision. We don't know as much about what is going on inside the sun as we had thought. Does it mean that a catastrophe is on the way?

We can't say that. As far as our observation is concerned the sun has been stable enough throughout the history of life to make life continually possible on the planet. We had a theory that would account for the stability. Now we may have to modify that theory, but the modified theory will still have to account for the stability. The sun won't suddenly become unstable just because we have to modify our theory.

To summarize, then: A catastrophe of the second class, involving changes in the sun that will make life on Earth impossible, must come in no more than 7 billion years, but it will come with plenty of warning.

Catastrophies of the second class may come before then, and unexpectedly, but the chances of that are so small that it makes no sense to devote much time to worrying about it.

Part III

Catastrophes of the Third Class

Chapter 7

The Bombardment
of the Earth

Extraterrestrial Objects

In discussing the invasion of the solar system by objects from interstellar space earlier, I concentrated on the possibility that such objects might affect the sun, since any severe interference with the
integrity or properties of the sun is bound to have a fatal effect on us.

Still more sensitive than the sun to such misadventure is the Earth
itself. An interstellar object, passing through the solar system, might
be too small to affect the sun significantly barring a direct collision,
and sometimes not even then. Yet that same object, if it invaded the
neighborhood of the Earth, or collided with it, might bring on a catastrophe.

It is time, then, to consider catastrophes of the third class—those
possible events that will affect the Earth primarily and render it uninhabitable, though the universe, and even the rest of the solar system, remain unaffected.

Consider, for instance, the case of an invading mini-black hole of comparatively large dimensions—say, with a mass comparable to that of the Earth. Such an object, if it misses the sun, will do that body no harm, though it, itself, will perhaps will have its orbit drastically changed by the sun's gravitational field.*

If such an object skimmed by Earth, however, it could produce disastrous effects, even without making direct contact, entirely because of the influence of its gravitational field upon us.

Since the intensity of a gravitational field varies with distance, that side of the Earth facing the intruder will be more strongly attracted than the side that is turned away from the intruder. The Earth will be stretched to some extent in the direction of the intruder. In particular, the yielding waters of the ocean will be stretched. The ocean will hump up on opposite sides of the Earth, toward the intruder and away from it, and, as the Earth turns, the continents will pass through those humps. Twice a day, the sea will creep up the continental shores then recede again.

The advance and recession of the sea (the "tides") is actually experienced on the Earth as a result of the gravitational influence of the moon and, to a lesser extent, of the sun. It is because of this that all effects produced by differences in gravitational influence on a body are called "tidal effects."

The tidal effects are greater, the larger the mass of the intruder and the closer it passes to Earth. If an invading mini-black hole is massive enough and skims by Earth closely enough, it might actually interfere with the integrity of the planetary structure, produce breaks in its crust, and so on. An actual collision would, of course, be clearly catastrophic.

Such a sizable mini-black hole would be exceedingly rare, however, even if it existed at all, and we must remember that the Earth is a much smaller target than the sun is. The cross-sectional area of the Earth is only a twelve-thousandth that of the sun, so that the very small chance that there would be a close encounter between such a body and the sun must be further decreased by a factor of twelve thousand for a close encounter with the Earth.

Mini-black holes, if they exist at all, would be much more likely to

---

* It might even (though that is not likely) be captured by the sun and go into permanent orbit about it. That orbit is likely to be highly inclined to the ecliptic and highly eccentric. With luck, it would not disturb the other bodies of the solar system, including Earth, appreciably, though it would be, and remain, a most uncomfortable neighbor. It is very unlikely that a large mini-black hole is a member of the solar system, though. The tiny effects of its gravitational field would have been noticed, unless it lies a substantial distance beyond Pluto's orbit.

be of asteroidal size. An asteroidal mini-black hole, with a mass that was, say, only a millionth that of the Earth, would offer no serious dangers in a near-miss. It would produce insignificant tidal effects and we might well be unaware of such an event if it took place.

It would be different, however, in case of a direct hit. A mini-black hole, however small, would tunnel its way into the Earth's crust. It would absorb matter, of course, and the energies given off in the process would melt and vaporize the matter ahead of it in its path. It could tunnel all the way through in a curved line (though, of course, not necessarily passing through the center) and emerge from Earth to continue its path through space—one that was altered by Earth's gravitational pull, of course. It would be more massive when it emerged than it had been when it entered. It would also be moving more slowly, for in passing through the gases of Earth's vaporizing substance, it would have encountered a certain resistance.

The body of the Earth would heal itself after the mini-black hole had passed on its way. The vapors would cool and solidify, and internal pressures would close up the tunnel. The effect on the surface would, however, be that of an enormous explosion—two of them, in fact, one in the area where the mini-black hole entered, and one where it emerged—with devastating (though perhaps not completely catastrophic) effects.

Naturally, the smaller the mini-black hole, the smaller the effects, except that in one respect a small one may actually be worse than a rather larger one. A small mini-black hole would have a rather low momentum, thanks to its small mass, and if it also happened to be moving at a low speed relative to Earth, it might be just sufficiently slowed in the process of tunneling through to be unable to work its way out the other end. It would then be trapped in Earth's gravity. It would fall toward the center, overshoot the mark, fall back again and so on, over and over.

Because of the Earth's rotation, it would not go back and forth in the same track, but would rather carve out an intricate honeycomb of tracks, growing steadily larger as it did so and absorbing more matter in each sweep. Eventually, it would settle down at the center, leaving behind a riddled Earth, with a hollowed-out region at the center— that hollow slowly growing. If the Earth were so weakened structurally in this fashion that it would collapse, more material would make its way into the central black hole and eventually, the entire planet might be consumed.

The resulting black hole, with the mass of Earth, would continue to move in Earth's orbit around the sun. To the sun and to the other planets it would make no gravitational difference whatever. Even the moon would continue to sweep around a tiny object, 2 centimeters (0.8 inches) across, just as though it were the full-sized Earth, which, from the standpoint of mass, it would be, of course.

It would, however, be the end of the world for us—the epitome of a catastrophe of the third class. And (in theory) it could happen tomorrow.

Again, a piece of antimatter, too small to disturb the sun appreciably, even if it collided with that body, might well be large enough to wreak havoc with the Earth. Unlike the black hole, it would not, if it were of asteroidal mass or less, tunnel right through the planet. It would, however, gouge out a crater that could destroy a city, or a continent, depending on its size. Ordinary chunks of matter of the familiar variety, invading from interstellar space, would naturally do even less damage.

Earth is protected from these catastrophes by invasion for two reasons:

1. In the case of mini-black holes and of antimatter bodies, we don't really know that objects of that sort exist at all.

2. If these objects did exist, space is so huge in volume and the Earth is so small a target that it would take the most extraordinary fall of all-but-impossible odds for us to be struck, or even to undergo a close approach. This would hold for objects of ordinary matter, too, of course.

On the whole, then, we might dismiss invaders from interstellar space, sizable invaders of any kind, as representing no perceptible danger to Earth.*

Comets

If we were to seek missiles that might be launched against the Earth we need not seek invaders from interstellar space. There are objects to spare in the solar system itself.

It has been well known since about 1800, thanks to the work of the French astronomer Pierre Simon Laplace (1749–1827) that the solar system is a stable structure, provided that it is left to itself. (And it

---

* By saying "sizable," I am deliberately omitting the possibility of collision with the Earth of dust particles from interstellar space, or of individual atoms or subatomic particles. I will consider these later.

has been left to itself, as far as we know, for the 5 billion years of its existence and should be left to itself, as far as we can judge, for an indefinite period to come.)

For instance, the Earth cannot fall into the sun. In order for it to do so it would have to get rid of its enormous supply of angular momentum of revolution. That supply cannot be destroyed; it can only be transferred; and we know of no mechanism short of the invasion of a planet-sized body from interstellar space that would absorb the Earth's angular momentum, leaving Earth motionless with respect to the sun and therefore capable of falling into it.

For the same reason no other planet can fall into the sun, no satellite can fall into its planet, and, in particular, the moon cannot fall into the Earth. Nor can planets so alter their orbits as to collide with each other.*

The solar system was, of course, not always as orderly as it now is. When the planets were f0irst forming, a cloud of dust and gas on the outskirts of the coalescing sun condensed into fragments of varying sizes. The larger fragments grew at the expense of the smaller until large cores of planetary size were formed. There still remained smaller objects of considerable size, however. Some of them became satellites, circling the planets in what came to be stable orbits. Others actually collided with the planet or the satellites, and added the last bits of mass to them.

We can see the marks of the final collisions with the moon, for instance, using nothing more than a good pair of binoculars. There are 30,000 craters on the moon with diameters ranging from 1 kilometer to over 200 kilometers—each the mark of a collision of a speeding bit of matter.

Rocket probes have shown us the surfaces of other worlds, and we find craters on Mars, on both of its two small satellites, Phobos and Deimos, and on Mercury. The surface of Venus is cloud-covered and hard to explore but there are doubtless craters there, too. There are craters even on Ganymede and Callista, two of the satellites of Jupiter. Why is it, then, that there are no bombardment craters on Earth?

---

* To be sure, the Russian-born psychiatrist Immanuel Velikovsky (1895–) in his book *Worlds in Collision*, published in 1952, postulated a situation in which the planet Venus had been spewed out of Jupiter, about 1500 B.C., and had then had several encounters with Earth before settling down into its present orbit. Velikovsky describes a number of disastrous events following these encounters which, however, seem to have left no mark on the Earth, if one doesn't count the vague myths and folk-tales which Velikovsky selectively quotes. Velikovsky's ideas can safely be dismissed as fantasies born of an active imagination that appeal to people whose knowledge of astronomy is no greater than Velikovsky's.

Oh, but there are! Or, rather, there once were. Earth has characteristics which other worlds its size lack. It has an active atmosphere, which the moon, Mercury, and the Jovian satellites lack, and which Mars possesses in only a small amount. It has a voluminous ocean, to say nothing of ice, rain, and running water, and this no other object shares, although there is ice and there may once have been running water on Mars. Finally, Earth has life, something in which it appears to be unique in the solar system. Wind, water, and life-activity all serve to erode surface features, and since the craters were formed billions of years ago, those on Earth are now erased.*

Within the first billion years after the formation of the sun, the various planets and satellites had swept their orbits clear and had taken on their present shape. Yet the solar system is not entirely clear even now. There remains what we might call planetary debris, small objects, circling the sun, that are far too small to make respectable planets, and yet that are capable of creating considerable havoc if they somehow collided with a larger body. There are, for instance, the comets.

Comets are hazy objects, glowing fuzzily and sometimes having irregular shapes. They have been seen in the sky for as long as human beings have looked at the sky, but their nature was unknown till modern times. The Greek astronomers thought they were atmospheric phenomena and consisted of burning vapors high in the air.†
It was not till 1577 that the Danish astronomer, Tycho Brahe (1546–1601) was able to show that they existed far out in space and must wander among the planets.

In 1705, Edmund Halley was finally able to calculate the orbit of a comet (Halley's comet, it is now called). He showed that the comet did not move around the sun in a nearly circular orbit as the planets did, but in an enormously elongated ellipse of high eccentricity. It was an orbit that brought it comparatively near the sun at one end and took it far beyond the orbit of the farthest known planets at the other.

The fact that comets visible to the naked eye have an extended appearance instead of being mere points of light, as are the planets

---

* Recent photographs of Io, the innermost of Jupiter's large satellites, show it to be crater-free. In its case, the reason is that it is actively volcanic and the craters are obscured by lava and by ash.
† Because comets appeared according to no regular rule, as contrasted with the steady and predictable movements of the planets, they seemed to most people of prescientific ages to be portents of disaster created specially and sent as warnings to humanity by angry gods. It was only gradually that scientific investigation allayed these superstitious fears. In fact, those fears are not completely gone even yet.

and stars, made it look as though they might be very massive bodies. The French naturalist George L. L. Buffon (1707–88) thought this was so, and considering the manner in which they seemed to skim by the sun at one end of their orbit, wondered if by a slight miscalculation, so to speak, one might actually hit the sun. In 1745, he suggested that it might be through such a collision that the solar system was formed.

Nowadays we know that comets are actually small bodies, not more than a few kilometers across at most. According to some astronomers, such as the Dutch astronomer Jan Hendrik Oort (1900–), there may be as many as a hundred billion of these bodies forming a shell around the sun at a distance of a light-year or so. (Each of these would be so small and all of them would be scattered over so huge a volume of space that they would not interfere with our view of the universe at all.)

The comets may well be unchanged residues of the outskirts of the original cloud of dust and gas out of which the solar system was formed. They are probably made out of compounds of the lighter elements, frozen as icy substances—water, ammonia, hydrogen sulfide, hydrogen cyanide, cyanogen, and so on. Embedded in these ices would be various quantities of rocky material in the form of dust or gravel. In some cases, rock may form a solid core.

Every once in a while a comet of this far-distant shell may be perturbed by the gravitational influence of some comparatively nearby star and may take up a new orbit which will bring it closer to the sun; sometimes very close to the sun. If, in passing through the planetary system, the comet is perturbed by the gravitational pull of one of the larger planets, its orbit may change again and it may remain within the planetary system until another planetary perturbation casts it out once more.*

When a comet swings into the inner solar system, the heat of the sun begins to melt the ice and a cloud of vapor, made visible by the inclusion of particles of ice and dust, envelops the comet's central "nucleus." The solar wind sweeps the cloud of vapor away from the sun and stretches it into a long tail. The larger and icier the comet and the closer it comes to the sun, the larger and brighter the tail. It is this cloud of dust and vapor, lengthened into a tail, that gives the

---

* Comets are small and consequently have far, far less mass and angular momentum than planets do. The tiny transfers of angular momentum through gravitational interaction, which produce immeasurably tiny orbital effects in the case of planets and satellites, are sufficient to alter cometary orbits, in some cases, drastically.

comet its huge apparent size, but it is a very insubstantial cloud and represents very little mass.

After a comet passes the sun and returns to the far reaches of the solar system, it is smaller by the amount of material it lost in the passage. It loses more with each additional visit to the neighborhood of the sun, until it dies altogether. It is either reduced to its central core of rock, or, if there is none, to a cloud of dust and gravel that slowly spreads throughout the cometary orbit.

Since the comets originate from a shell that surrounds the sun in three dimensions, they can enter the solar system at any angle. Since they are easily perturbed, their orbits can be almost any kind of ellipse, taking up any position with respect to the planets. In addition, the orbit is always subject to change with further perturbation.

Under these conditions, a comet is not as well-behaved a member of the solar system as the planets and satellites are. Any comet might, sooner or later, hit some planet or satellite. In particular, it might hit the Earth. What keeps it from happening is merely the vastness of space and the comparative smallness of the target. Nevertheless, there is an enormously better chance of the Earth's being hit by a comet than by any sizable object from interstellar space.

For instance, on June 30, 1908, on the Tunguska River in the Russian Empire—quite near the exact center of the empire, in fact—there was a huge explosion at 6:45 A.M. Every tree was knocked down for a score of miles in all directions. A herd of reindeer was wiped out and undoubtedly innumerable other animals were killed. Fortunately, not a single human being was hurt! The explosion took place in the midst of an impenetrable Siberian forest, and neither people nor the works of people were within the wide range of the devastation. It was not until years later that the site of the explosion could be investigated and it was then discovered there was no sign of any impact with the Earth. There seemed to be no crater, for instance.

Ever since, explanations have been offered to account for the violence of the event and the lack of impact—mini-black holes, antimatter, even extra-terrestrial spaceships with exploding nuclear engines. Astronomers, however, are reasonably sure that it was a small comet. The icy material that made it up evaporated as it plunged through the atmosphere, and so rapidly as to explode shatteringly. The explosion in the air, perhaps less than 10 kilometers (6 miles) above the ground, would do all the damage that the Tunguska explosion did in fact do, but the comet would, of course, never have reached the ground, so that there would naturally be no crater and no fragments of its structure strewn about the site.

It was pure good fortune that the explosion hit in one of the few places on Earth where no damage to human beings was done. In fact, if the comet had followed exactly the course it had taken but if the Earth had happened to have been one-quarter turn farther along in its rotation, the city of St. Petersburg (now Leningrad) would have been wiped out. We were lucky that time, but it may happen again someday with worse effects, and we don't know when. There is not likely to be any warning under present conditions.

If we count the comet's tail as part of the comet, then the possibility of collision becomes greater still. Cometary tails can stretch for many millions of kilometers and occupy so great a volume of space that the Earth might easily move through one. Indeed, in 1910, the Earth did pass through the tail of Halley's comet.

Cometary tails, however, represent matter so thinly spread out that they are very little better than the vacuum of interplanetary space itself. Though composed of poisonous gases that could be dangerous if the tail were as dense as Earth's atmosphere, at their own typical density they are harmless. Earth suffered no noticeable effect, none whatever, in passing through the tail of Halley's comet.

The Earth may also pass through the dusty material left over by dead comets. Indeed, it does. These dust specks are constantly striking the Earth's atmosphere and slowly settle to Earth, serving as nuclei for rain drops. Most are microscopic in size. Those that are of visible size heat up as they compress the air before them and give off light, shining as a "shooting star" or "meteor" until vaporized.

None of these objects can do any damage, but merely settle to the ground eventually. Although so small, so many of them strike the Earth's atmosphere that it is estimated that the Earth gains about 100,000 tons of mass from these "micrometeoroids" each year. This sounds like a great deal but in the last 4 billion years such an access of mass, if it had kept up steadily at that rate, would amount to less than 1/10,000,000 the total mass of the Earth.

Asteroids

The comets are not the only small bodies of the solar system. On January 1, 1801, the Italian astronomer Giuseppi Piazzi (1746–1826) discovered a new planet which he named Ceres. It moved around the sun in a typical planetary orbit, one that was nearly circular. Its orbit lay between those of Mars and Jupiter.

The reason it was not discovered until so late in history rested in the fact that it is a very small planet and therefore caught and reflected so little sunlight that it was too dim to be seen by the naked eye. It was, in fact, only 1,000 kilometers (600 miles) in diameter, considerably smaller than Mercury, the smallest planet known till that time. For that matter, it is smaller than ten of the satellites of the various planets.

If that were all, it would simply have been accepted as a pygmy planet, but there was more to it than that. Within six years of the discovery of Ceres, astronomers discovered three more planets, each even smaller than Ceres, and each with an orbit between those of Mars and Jupiter.

Since these new planets were so small, they appeared as merely starlike points of light in the telescope and were not expanded into discs as the planets themselves were. William Herschel therefore suggested the new bodies be called "asteroids" ("starlike") and the suggestion was adopted.

As time passed, more and more asteroids were discovered, all of them either smaller or farther from Earth (or both) than the first four, and therefore still dimmer and harder to see. By now, well over 1,700 asteroids have been located and have had their orbits calculated. It is estimated that there are anywhere from 40,000 to 100,000 altogether with diameters of more than a kilometer or so. (Again, they are individually so small and are scattered over so huge a volume of space that they do not interfere with the astronomers' view of the sky.)

Asteroids differ from comets in being rocky or metallic rather than icy. The asteroids can be considerably larger than comets, too. Asteroids can therefore be, at their worst, more formidable projectiles than comets are.

Asteroids, however, are for the most part in more secure orbits. Almost all asteroidal orbits lie through all their length in that portion of planetary space between the orbits of Mars and Jupiter. If all of them remained there permanently, they would, of course, represent no danger to Earth.

Asteroids, however, particularly the smaller ones, are subject to perturbations and orbital changes. In the course of time, some orbits change in such a way as to carry asteroids particularly close to the limits of the "asteroid belt." At least eight of them came close enough to Jupiter itself to be captured and are now satellites of that planet, circling it in distant orbits. There may be other such satellites of Jupiter that are too small to have yet been detected. Then, too, there are several dozen satellites which, while not captured by Jupiter itself, travel in Jupiter's orbit either 60 degrees ahead of it or 60 degrees behind it, locked more or less in place by Jupiter's gravitational influence.

There are even asteroids whose orbits have been perturbed into elongated ellipses such that when the asteroids are nearest the sun they are in the asteroid belt, but at the other end of their orbit move well beyond Jupiter. One such asteroid, Hidalgo, discovered in 1920 by the German astronomer Walter Baade (1893–1960) moves out nearly as far as the orbit of Saturn.

However, if the asteroids that stay within the asteroid belt are no danger to Earth, certainly those that stray beyond the outer limits of the belt and move beyond Jupiter are no danger either. But are there asteroids that stray in the other direction and move in within the orbit of Mars and possibly approach Earth?

The first indication of such a possibility came in 1877 when the American astronomer Asaph Hall (1829–1907) discovered the two satellites of Mars. They were tiny objects of asteroidal size and they are now thought to be captured asteroids that ventured too close to Mars. Then, on August 13, 1898, the German astronomer Gustav Witt discovered an asteroid he named Eros. Its orbit was markedly elliptical in such a way that when it was farthest from the sun it was well within the asteroid belt, but when it was nearest the sun, it was only 170 million kilometers (106 million miles) from the sun. That brings it almost as close to the sun as Earth is.

In fact, if both Eros and Earth were at the proper points in their orbits, the approach would be only 22.5 million kilometers (14 million miles) apart. Naturally, it is not often that both are in the appropriate points in their orbits and usually they are considerably farther apart than that. Nevertheless, Eros can approach Earth more closely than any planet can. It was the first sizable object in the solar system (other than the moon itself) ever found to approach Earth more closely than Venus does, and therefore is considered the first of the "Earth-grazers" to have been spotted.

In the course of the twentieth century, as photography and other techniques were used to detect asteroids, over a dozen other Earth-grazers were discovered. Eros is an irregularly shaped object, with its longest diameter about 24 kilometers (15 miles), but the other Earth-grazers are all smaller than this, most having diameters of from 1 to 3 kilometers.

How close can an Earth-grazer get? In November, 1937, an asteroid, which was given the name of Hermes, was observed to streak past the Earth at a distance of no more than 800,000 kilometers (500,000 miles) scarcely twice the distance of the moon. An orbit calculated for it at the time showed that if Hermes and Earth were at the proper points in their orbit the approach would be as close as 310,000 kilometers (190,000 miles) and at such a time, Hermes would be even closer to us than is the moon. This is not a comfortable thought, for Hermes is probably a kilometer across and a collision with it would do enormous damage.

We can't be certain about the orbit, though, for Hermes has never been sighted again, which means that the orbit as calculated was not correct or that Hermes was perturbed out of that orbit. If it is sighted again, it will be only by accident.

Of course, there are undoubtedly many more Earth-grazers in existence than we are likely to see with our telescopes since any object passing Earth at a close distance does so very rapidly and may be missed altogether. Then, too, if it were very small (and, as in all such cases, there are more small Earth-grazers than large ones) it would be very dim even at best.

The American astronomer Fred Whipple (1911–) suspects there may be at least 100 Earth-grazers larger than 1.5 kilometers in diameter. It follows from that that there may well be some thousands of additional ones that are between 1.5 and 0.1 kilometers in diameter.

On August 10, 1972, a very small Earth-grazer actually passed through the upper atmosphere and in the process was heated to a visible glow. At its closest approach it was 50 kilometers (30 miles) over southern Montana. Its diameter is estimated to be 0.013 kilometers (14 yards).

In short, the region in the neighborhood of Earth seems to be rich in objects that nobody had ever seen prior to the twentieth century, from an object as huge as Eros, down through dozens of objects that are the size of mountains, thousands of objects that are the size of large boulders, and billions of objects that are pebbles. (If we want to count cometary debris, which I have already mentioned in the previous section, there are uncounted trillions of objects that are of pinhead size and less.)

Can the Earth pass through so populated a space and undergo no collisions? Of course not. Collisions take place constantly.

Meteorites

In almost all cases, those fragments of matter large enough to be heated to a visible glow as they streak through the atmosphere (at which time they are called "meteors") are vaporized to dust and vapor long before they reach the ground. This is invariably true of cometary debris.

Perhaps the greatest "meteor shower" in historic times came in 1833 when, to observers in the eastern United States, the flashing streaks of light seemed as thick as snowflakes, and the less sophisticated thought the stars were falling out of the sky and the world was coming to an end. When the meteor shower was over, however, all the stars were still shining in the sky as serenely as ever. Not one was missing. What's more, not one of those flashing bits of matter struck the ground as an object of detectable size.

If a piece of debris striking the atmosphere is large enough, its rapid passage through the air does not suffice to vaporize it entirely, and a portion of it then reaches the ground as a "meteorite." Such objects are probably never of cometary origin but are small Earth-grazers which originated in the asteroid belt.

Perhaps 5,500 meteorites have hit the Earth's surface in historic times, and about one-tenth of them have been iron, while the remainder have been stone.

The stone meteorites, unless actually seen to fall, are difficult to distinguish from the ordinary rocks of Earth's surface for anyone but a specialist in such matters. The iron meteorites* are, however, very noticeable, since metallic iron does not occur naturally on Earth.

In the days before it was learned how to obtain iron by smelting iron ore, meteorites were a valued source of a super-hard metal for points and edges of tools and weapons—far more valuable than gold, if less pretty. So assiduously were they sought that no iron meteorite fragments have ever been found in modern times in those areas where civilization flourished before 1500 B.C. The pre-Iron Age cultures found and used them all.

* Actually, they are steel alloy, for they are mixed with nickel and cobalt.

The meteorite finds were not equated with meteors, however. Why should they be? A meteorite was just a piece of iron found on the ground; a meteor was a flashing light high in the air;* why should there be any connection?

To be sure there were legends of objects falling from the heavens. The "black stone" in the Kaaba, holy to Moslems, may be a meteorite that was seen to fall. The original object of veneration in the temple of Artemis at Ephesus may have been another. Scientists in early modern times dismissed such stories, however, and considered any tales of objects falling from the sky to be superstition.

In 1807, an American chemist at Yale, Benjamin Silliman (1779–1864), and a colleague reported witnessing a meteorite landing. President Thomas Jefferson, on hearing the report, stated that it was easier to believe that two Yankee professors would lie than that stones would fall from heaven. Nevertheless, scientific curiosity was aroused by continuing reports, and while Jefferson was being skeptical, the French physicist Jean Baptiste Biot (1774–1862) had already, in 1803, written a report on meteorites that led to the acceptance of such falls as a true phenomenon.

For the most part, meteorites that have fallen in civilized areas have been small and have done no particular damage. There is only one report of any human being having been struck by a meteorite and that involves a woman in Alabama who some years ago received a glancing blow and bruised her thigh.

The largest known meteorite is still in the ground in Namibia in southwest Africa. It is estimated to weigh about 66 tons. The largest known iron meteorite on display is at the Hayden Planetarium in New York and weighs about 34 tons.

Even meteorites no larger than that could do considerable damage to property and kill hundreds, even thousands, of people, if they landed in a densely populated city area. What are the chances, though, that a really large strike might take place someday? Out in space, there are some pretty big mountains on the loose, which could do enormous damage if they struck us.

We might argue that the big objects in space (which are much fewer than the small objects, of course) are in orbits that don't intersect that of the Earth and never come anywhere near us. That would

* "Meteor" is from a Greek word for "upper atmosphere," since to the ancient Greeks, meteors, like comets, seemed purely atmospheric phenomena. Thus it is that "meteorology" is the study of the weather, not meteors. The study of meteors in the modern sense is called "meteoritics."

explain why we haven't been really slammed before this and therefore why we need not fear a slam in the future.

This argument is, however, not reassuring for two reasons. In the first place, even if the large meteoric objects have orbits that do not intersect ours, future perturbations may alter those orbits and place the object on a potential collision course. Second, there *have* been fairly large strikes; large enough to destroy a city, let us say. And if they have not actually taken place in historic times, they have fallen not too long before, geologically speaking.

The evidence for such strikes is not easy to obtain. Imagine a big strike taking place some hundreds of thousands of years ago. The meteorite would probably have buried itself deep in the ground where it could not easily be recovered and studied. It would have left behind a large crater, to be sure, but the action of wind, water, and life would have eroded it away completely in a few thousand years.

Even so, signs have been discovered of round formations, sometimes filled or partly filled with water, which can be seen easily from the air. The roundness, combined with clear differences from surrounding formations, rouses keen suspicion of the presence of a "fossil crater" and closer observation may then confirm it. Perhaps twenty such fossil craters have been located here and there on Earth and these have probably all been formed within the last million years.

The largest fossil crater definitely identified is the Ungava-Quebec crater, in the Ungava peninsula that makes up the northernmost part of the Canadian province of Quebec. It was discovered in 1950 by Fred W. Chubb, a Canadian prospector (so that it is sometimes called Chubb crater), from aerial photographs that showed the existence of a circular lake surrounded by other, smaller, circular lakes. The crater is 3.34 kilometers (2.07 miles) in diameter and 0.361 kilometers (401 feet) deep. The rim of the lake stands 0.1 kilometers (330 feet) above the surrounding countryside.

Clearly, if a strike like that were to be repeated and were to fall on Manhattan, it would destroy the entire island, severely damage parts of neighboring Long Island and New Jersey, and kill several million people.

A smaller, but much better preserved crater is one located near the town of Winslow, Arizona. In that dry area, there has been no water and little in the way of life to erode the crater. It looks fresh even today, and seems remarkably like a small cousin of the kind of crater we see on the moon.

It was discovered in 1891, but the first person to insist that the crater was the result of a meteoritic impact, rather than being an extinct volcano, was Daniel Moreau Barringer in 1902. It is therefore called the "Great Barringer Meteor Crater," or sometimes just "Meteor Crater."

Meteor Crater is 1.2 kilometers (0.75 miles) across and about 0.18 kilometers (600 feet) deep. Its rim rises nearly 0.060 kilometers (200 feet) above the surrounding countryside. The crater may have been formed as long as 50,000 years ago, though some estimates as low as 5,000 years have been offered. The weight of the meteorite that produced the crater has been estimated by various people as low as 12,000 tons and as high as 1.2 million tons. This means that the meteorite may have been anywhere from 0.075 to 0.360 kilometers (250 to 1,200 feet) in diameter.

But this is all in the past. What may we expect in the future? The astronomer Ernst Öpik estimates that an Earth-grazer should travel in its orbit for an average of 100 million years before colliding with Earth. If we suppose that there are two thousand such objects large enough to wipe out a city or worse, if they strike, then the average interval of time between such calamities is only 50,000 years.

What are the chances of a particular target being hit—say New York City? The area of New York City is 1.5 millionths of the Earth's area. That means the average interval between strikes that might destroy New York City is about 33 billion years. If we assume that the total area of large-city populations on Earth is a hundred times that of New York City, the average interval between city-destroying strikes somewhere on Earth is 330 million years.

This is not really something to lose sleep over, and it is not surprising that in the established written records of human civilization (which is only five thousand years old) there is no clear description of a city being destroyed by a falling meteorite.*

A sizable meteorite need not strike a city directly to do enormous damage. It one hit the ocean, which on the basis of chance seven out of ten meteorites will do, a tidal wave would be set up that would ravage coastlines, drowning people and destroying the works of man. If the average time between strikes is 50,000 years, then the average time between meteorite-induced tidal waves should be 71,000 years.

The worst of it is, of course, that there is, as of now, no possibility of advance warning of any meteor strike. The colliding object would quite likely be small enough and be moving quickly enough to reach

---

* It may, of course, be possible that the tale of the destruction of Sodom and Gomorrah, as described in the Bible, is the dim and distorted memory of a meteor strike.

Earth's atmosphere unnoticed. By the time it began to glow, it would be only a matter of minutes before the strike.

If the devastation of a large meteor strike is somewhat less unlikely than any of the other catastrophes I have discussed so far, it differs from them in two ways. In the first place, though it may be disastrous and do untold damage, it is not at all likely to be catastrophic in the sense that the sun's becoming a red giant would be. A meteorite is not likely to destroy the Earth or to wipe out humanity or even to topple our civilization. In the second place, it may not be long before this particular type of disaster may become completely preventable, even before the first disastrous strike of the future occurs.

We are moving out into space and within the century there may be elaborate astronomical observatories on the moon and in orbit about the Earth. Without an interfering atmosphere, astronomers at such observatories will have a better chance to sight the Earth-grazers. They can watch those dangerous bodies more closely and plot their orbits more carefully. This will include those Earth-grazers that are too small to be seen from Earth's surface, but that are still large enough to destroy a city and that, because of their greater numbers, are far more dangerous than the real giants.

Then perhaps, a hundred years from now, or a thousand, some space astronomer will look up from his computer to say, "Close-encounter orbit!" And a counterattack, kept in waiting for this necessary moment for decades or centuries, would be set in motion. The dangerous rock would be stalked and, at a convenient, precalculated position in space, some powerful device would be sent to intersect it and blow it up. The rock would glow and vaporize and change from a boulder to a conglomeration of pebbles. Earth would avoid the damage and, at worst, be treated instead to a spectacular meteor shower.

Eventually, perhaps every object that showed even the slightest potential for coming too close, and which astronomers would certify as having no further scientific value, would be blasted. This particular type of disaster would then nevermore need to concern us.

Chapter 8

The Slowing
of the Earth

Tides

As I have said, the chance of a catastrophe of the third class—the destruction of the Earth as an abode for life by some process that does not involve the sun—through invasion from space beyond the orbit of the moon is not something to be concerned with. It is either very unlikely or is not truly catastrophic or is, in some cases, on the verge of being preventable. We must next ask ourselves, however, if there is anything that can produce a catastrophe of the third class that does not involve objects from beyond the Earth-moon system. To begin with, then, we must consider the moon itself.

The moon is by far the closest to Earth of any sizable astronomical body. The distance from the moon to the Earth, center to center, is 384,404 kilometers (238,868 miles). If the moon's orbit about the Earth were perfectly circular, this would be its distance at all times.

150

The orbit is, however, slightly elliptical, which means that the moon can come as close as 356,394 kilometers (221,463 miles) and can recede as far as 406,678 kilometers (252,710 miles).

The moon is at only 1/100 the distance of Venus, when the latter body is at its closest to Earth; only 1/140 the distance of Mars at its closest, and only 1/390 the distance of the sun at its closest. No object larger than the once-observed asteroid Hermes, which is certainly not more than a kilometer across, has approached Earth anything like as closely as the moon does.

To indicate the moon's closeness in another way, it is the only astronomical body close enough (so far) for human beings to reach, so we can say it is three days away from us. It takes about as long to reach the moon by rocket as it does to span the United States by rail.

Is the moon's extraordinary closeness in itself a danger? Might it fall for some reason and strike the Earth? If it did, it would be a far worse calamity than any collision with an asteroid, for the moon is a sizable body indeed. Its diameter is 3,476 kilometers (2,160 miles) or a little over a quarter that of Earth. Its mass is 1/81 that of the Earth and 50 times that of the largest asteroid.

If the moon fell to Earth, the consequences of the collision would certainly be fatal to all life on our planet. Both objects might be smashed and broken in the process. Fortunately, as I said in passing in the previous chapter, there is no chance whatever of this happening except as part of an even greater catastrophe. The moon's angular momentum cannot be removed suddenly and completely, so that it will fall in the usual sense of the word, except through transfer to some sizable third body approaching closely enough from just the right direction and at just the right speed. The chances of that happening are entirely negligible, so that we may cancel any fears that the moon won't stay in its orbit.

Nor need we fear that anything can happen to the moon that will involve only itself and that will contain the seeds of catastrophe for Earth. There is no chance at all, for instance, that the moon will explode and that we will be showered by its fragments. The moon is, geologically, just about dead and its internal heat is not sufficient to produce any effects that will noticeably change its structure or even its surface.

In fact, we can safely assume that the moon will remain very much as it is today, barring exceedingly slow changes, and that its corporeal body will offer us no danger until such time as the sun expands into a red giant and both moon and Earth are destroyed.

The moon does not have to strike the Earth with all or part of itself to affect us, however. It exerts a gravitational influence across space, and that gravitational influence is a strong one. It is, in fact, second only to that of the sun.

The gravitational influence of any astronomical object upon Earth depends on the mass of that object, and the sun has a mass that is 27 million times that of the moon. The gravitational influence also decreases as the square of the distance, however. The sun's distance from Earth is 390 times that of the moon, and 390 × 390 = 152,100. If we divide this into 27,000,000, we find that the sun's gravitational attraction on Earth is 178 times that of the moon's gravitational attraction upon us.

Although the moon's pull on us is only 0.56 percent that of the sun, that is far larger than any other gravitational pull upon us. The moon's pull upon us is 106 times greater than that of Jupiter upon us when it is at its closest and 167 times that of Venus at its closest. The gravitational pull on Earth of astronomical objects other than Venus and Jupiter is far smaller still.

Can the moon's gravitational attraction on us be the seed of catastrophe, then, when it is so large compared to all objects other than the sun? The answer might seem, at first glance, to be in the negative, since the gravitational pull of the sun is much greater than that of the moon. Since the former causes no trouble, why should the latter?

This would be true if astronomical objects reacted to gravitational pulls equally at all points—but they don't. Let us return to the matter of tidal effects, which I mentioned briefly in the previous chapter, and consider it in greater detail with respect to the moon.

The surface of the Earth facing the moon is at an average distance of 378,026 kilometers (234,905 miles) from the center of the moon. The surface of the Earth facing away from the moon is farther from the moon's center by the thickness of the Earth, and is therefore 390,782 kilometers (242,832 miles) away.

The strength of the moon's pull decreases as the square of the distance. If the distance of the Earth's center from the moon's center is called 1, then the distance of the Earth's surface directly facing the moon is 0.983 and the distance of the Earth's surface facing directly away from the moon is 1.017.

If the gravitational pull of the moon on Earth's center is set at 1, then the pull on Earth's surface facing the moon is 1.034, and the pull on Earth's surface away from the moon is 0.966. This means that the moon's pull on the Earth's near surface is 7 percent greater than on its far surface.

The result of the moon's pull on Earth changing with distance in this way is that the Earth is stretched in the direction of the moon. The side toward the moon is pulled harder than the center is, and the center is, in turn, pulled harder than the side away from the moon.

As a result, the Earth bulges on each side. One bulge is toward the moon, pushing toward the moon more eagerly than the rest of Earth's structure, so to speak. The other bulge is on the side away from the moon, lagging behind the rest, so to speak.

Since the Earth is made of stiff rock that doesn't yield much even to hard pulls, the bulge in Earth's solid body is small but it is there. The water of the ocean is, however, more yielding, and forms a larger bulge.

As the Earth turns, the continents pass through the higher bulge of water facing the moon. The water creeps some distance up the shoreline and then recedes again—a high tide and a low tide. On the other side of the Earth, facing away from the moon, the turning continents pass through the other bulge about 12½ hours later (the extra half hour arises from the fact that the moon has moved somewhat in the interval). There are thus two high tides and two low tides a day.

As it happens, the tidal effect produced on the Earth by any body is proportionate to its mass, but decreases as the *cube* of its distance. The sun (to repeat) is 27 million times as massive as the moon and is 390 times as far away. The cube of 390 is just about 59,300,000. If we divide the sun's mass (relative to the moon) by the cube of its distance (relative to the moon) we find that the tidal effect of the sun upon the Earth is 0.46 times that of the moon.

We conclude, then, that the moon is the major contributor to tidal effects on the Earth and the sun a minor contributor. All other astronomical bodies have no measurable tidal effect on the Earth at all.

Now we must ask whether the existence of tides can, in any way, presage a catastrophe.

The Longer Day

To speak of tides and catastrophes in one breath seems odd. There have always been tides throughout human history and they have been perfectly regular and predictable. What's more they have been useful, since ships usually sailed at the turn of high tide, when the water lifted them high above any hidden obstacles and the retreating water pushed the ship in the direction it wanted to go.

What's more, tides can become useful in the future in another way. At high tide, water could be lifted into a reservoir from which it could emerge, as the tide lowered, to turn a turbine. Tides could, in this way, provide the world with an unending store of electricity. Where does the catastrophe come in?

Well, as the Earth turns, and as the dry land passes through the watery bulge, the water moving up and down the shore must overcome frictional resistance in doing so, not only from the shore itself but from those portions of the sea bottom where the ocean happens to be particularly shallow. Part of the energy of Earth's rotation is consumed in overcoming this friction.

What's more, as the Earth turns, the solid body of the planet also bulges, though only about a third as much as the ocean does. However, the Earth's bulge is produced at the expense of rock sliding against rock as the crust is pulled upward and released over and over. Part of the energy of Earth's rotation is consumed in this way, too. Of course, the energy isn't truly consumed. It doesn't disappear, but turns into heat. In other words, as a result of the tides, the Earth gains a little heat and loses a little of its speed of rotation. The day grows longer.

The Earth is so massive and it turns so quickly that it has an enormous store of energy. Even if a great deal of it (on a human scale) is consumed and turned to heat in overcoming tidal friction, the day would grow longer only very slightly indeed. Even a very slight increase in the length of the day would have a considerable cumulative effect however.

For instance, suppose the day started at its present length of 86,400 seconds and that it was, on the average, 1 second longer each year than it was the year before. At the end of a hundred years, the day would be 100 seconds or 1⅓ minutes longer. One would scarcely be aware of the difference.

Suppose, though, that you started the century with a watch that kept perfect time. By the second year it would be gaining 1 second each day compared to the sun; by the third year it would be gaining 2 seconds each day; by the fourth year it would be gaining 3 seconds each day and so on. By the end of the century, when the number of days would be 36,524, if we went by sunrises and sunsets, the clock would have recorded 36,534.8 sets of 86,400-second days. In short, by only increasing the length of the day by 1 second a year, we accumulate an error of almost 11 hours in a mere century. Of course, the day actually grows longer at a much slower rate. In ancient times, certain eclipses were recorded as taking place at a certain time of day. Calculating backward, we realize that they should have taken place at another time. The discrepancy is the accumulated result of a very slow lengthening of the day.

It might be argued that ancient people had only the most primitive methods for keeping time and that their whole concept of time recording was different from ours. It would therefore be risky to deduce anything from what they said about the time of eclipses.

It is not only time that counts, however. A total eclipse of the sun can be seen only from a small area of the Earth. If, let us say, an eclipse were to take place only one hour before the calculated time, the Earth would have had less time to turn and, in the temperate zone, the eclipse would have taken place perhaps 1,200 kilometers (750 miles) farther east than our calculations would indicate.

Even if we don't completely trust what ancient people have said about the time of an eclipse, we can be sure that they report the *place* of the eclipse accurately and that will tell us what we want to know. From their reports, we know the amount of the cumulative error, and, from that, the rate of the lengthening of the day. That is how we know that the Earth's day is increasing at the rate of 1 second every 62,500 years.

This seems anything but catastrophic. The day is now about 1/14 of a second longer than it was when the pyramids were built. Surely, such a discrepancy is small enough to be ignored. Surely! But historic times are an instant in comparison to geologic eras. The gain is 16 seconds in a million years and there are many million years in Earth's history.

Suppose we consider the situation as it was 400 million years ago when life, which had existed in the sea for nearly 3 billion years was finally beginning to emerge on land. In the last 400 million years, the day would have gained 6,400 seconds, if the present rate of increase had held through all that time.

The day 400 million years ago would then have been 6,400 seconds shorter than it is now. Since 6,400 seconds is equal to almost 1.8 hours, life would have crawled out onto land in a world in which the day was only 22.2 hours long. Since there is no reason to suppose that the length of the year has changed in that interval, it would also mean that there were 395 of those shorter days in one year.

This is only calculation. Can we find direct evidence? It seems there are fossil corals that date from about 400 million years ago. Such corals grow at a rate during the day that is different from the rate of growth at night; and at one rate in the summer and another in the winter. As a result, there are markings on the shell, rather like tree rings, that measure both the days and the years.

In 1963, the American paleontologist John West Wells studied those fossil corals carefully and found some 400 fine markings to every coarser marking. This would indicate that there were about 400 days to the year in those ancient times of 400 million years ago. That would mean that each day was 21.9 hours long.

That is quite close to the calculation. It is surprisingly close, in fact, for there is reason to think that the rate of increase of the day (or decrease if one goes back in time) is not necessarily constant. There are factors which change the rate of loss of rotational energy. The distance of the moon (as we shall soon see) changes with time; and so do the configuration of the continents, the shallowness of the seas, and so on.

Nevertheless, suppose (just for fun) that the day has been lengthening at a constant rate all through the history of the Earth. In that case, how rapidly was the Earth rotating 4.6 billion years ago when it was first formed? It is easy to calculate that allowing for a constant change in the length of day, Earth's period of rotation at its birth must have been 3.6 hours.

That is, of course, not necessarily so. More sophisticated calculations indicate that the day at its shortest may have been 5 hours long. Then, too, it may be that the moon did not accompany the Earth from the very beginning; that it was somehow captured at some period after Earth's formation, and that the tidal slowing began more recently than 4.6 billion years ago; perhaps considerably more recently. In that case, the day may have been 10 hours long or even 15 hours long in Earth's early days.

As yet, we can't be sure. We have no direct evidence for daylengths very early in Earth's history.

In any case, a shorter day in the far past is, in itself, of no great importance to life. A particular spot on Earth would have less time to warm up during a short day; less time to cool off during a short night. The temperatures of the primitive Earth would therefore tend to be somewhat more equable than they are now and it is quite obvious that living organisms could, and did, live with that. In fact, conditions may have been more favorable to life then than they are now.

What of the future, however, and of the continuing lengthening of the day?

The Receding Moon

As the millions of years pass, the day will continue to grow longer since the tides will not stop. Where will it end? We can get a notion of the end if we consider the moon, which is subjected to Earth's tidal influence as the Earth is subject to the moon's.

The Earth has 81 times the mass of the moon, so, if all things were equal, its tidal influence on the moon ought to be 81 times as great as the moon's is on us. All things are not equal, however. The moon is smaller than Earth, the distance across the moon being only a little over a quarter that of the distance across the Earth. For that reason, the Earth's gravitational pull undergoes a smaller drop from one side of the moon to the other and that decreases the tidal effect. Allowing for the moon's size, the Earth's tidal pull on the moon is 32.5 times that of the moon's on the Earth.

Still, that means that the moon is subjected to much greater frictional losses as it rotates, and since it has a considerably smaller mass than Earth has, it has less rotational energy to lose. The moon's rotational period must therefore have been lengthened at a much more rapid rate than was Earth's, and the moon's rotational period must now be quite long.

And so it is. The moon's period of rotation relative to the stars is now 27.3 days. This happens to be just equal to the period of revolution around the Earth relative to the stars so that the moon always presents the same face to the Earth as it revolves.

This is no accident, no wild coincidence. The moon's period of rotation slowed until it was slow enough to present the same face to the Earth at all times. Once that happened, the tidal bulge was always present at the same points on the moon's surface; one faced always toward the Earth from the side Earth always saw and one faced away from the Earth from the side Earth never saw. The moon no longer rotates relative to that tidal bulge and there is no more frictional conversion of energy of rotation into heat. The moon is gravitationally locked in place, so to speak.

If the Earth's rotation is slowing, then eventually it will rotate so slowly that it will always face one side to the moon and it, too, will be gravitationally locked in place.

Does this mean that the Earth will rotate so slowly that its days will be 27.3 present-days long? No, it will be worse than that for the following reason: You can turn the energy of rotation into heat, since that is a matter of turning one form of energy into another and doesn't violate the laws of conservation of energy. A rotating object, however, also has angular momentum and this cannot be turned into heat. It can only be transferred.

If we consider the Earth-moon system, the Earth and the moon each possess angular momentum for two reasons: Each rotates on its axis, and each revolves about a common center of gravity. The latter is located on the line connecting the center of the moon and the center of the Earth. If the Earth and moon were exactly equal in mass, the common center of gravity would be located just halfway between. Since the Earth is more massive than the moon, the common center of gravity is located closer to the Earth's center. In fact, since the Earth is 81 times as massive as the moon, the common center of gravity is 81 times as far from the moon's center as from the Earth's.

This means that the common center of gravity is located (if we consider the moon to be at its average distance from the Earth) 4,746 kilometers (2,949 miles) from the center of the Earth and 379,658 kilometers (235,919 miles) from the center of the moon. The common center of gravity is thus 1,632 kilometers (1,014 miles) below the surface of the Earth, on the side facing the moon.

While the moon makes a large ellipse about the common center of gravity every 27.3 days, the center of the Earth makes a much smaller ellipse about it in those same 27.3 days. The two bodies move in such a way that the moon's center and the Earth's center always stay on exactly opposite sides of the common center of gravity.

As the moon and the Earth each lengthen their rotational periods through the effect of tidal friction, each loses rotational angular momentum. To preserve the law of conservation of angular momentum, each must gain angular momentum associated with its revolution around the center of gravity in exact compensation of the loss of angular momentum associated with its rotation about its own axis. The way revolutionary angular momentum increases is for the Earth and moon to move farther away from the common center of gravity and thus make bigger swings about it.

In other words, as either the moon or the Earth, or both, lengthen their periods of rotation, they move away from each other and in this way the total angular momentum of the Earth-moon system remains the same.

In the far distant past, when the Earth spun more quickly on its axis and the moon had not yet slowed to the point of gravitational lock, the two bodies were closer together. If they had more rotational angular momentum, they had less revolutionary angular momentum. When the moon and Earth were closer together, they circled each other in less time, of course.

Thus, 400 million years ago, when the Earth's day was only 21.9 hours long, the distance from the moon's center to the Earth's was only 96 percent what it is now. The moon was only 370,000 kilometers (230,000 miles) from Earth. If we calculate backward in this fashion, it would appear that 4.6 billion years ago when the Earth was first formed, the moon was only 217,000 kilometers (135,000 miles) from Earth, or a little more than half its present distance.

The calculation is not a fair one, for as the moon comes closer to the Earth (as we look back in time), the tidal effect becomes greater, all things being equal. The chances are that early in Earth's history the moon was closer still, perhaps as close as 40,000 kilometers (25,000 miles).

Looking into the future, now, as the Earth's rotation period slows, the moon and the Earth will slowly separate. The moon is slowly spiraling away from the Earth. Each revolution about the Earth increases its average distance by about 2.5 millimeters (0.1 inches).

The moon's rotation will slow down very gradually so that it will continue to match the increasing length of the month. Eventually, when the Earth's period of rotation lengthens until it, too, faces one side to the moon at all times, the moon will have receded so far that the month will be 47 days long. At that time, the moon's rotation will be 47 present-days long as will that of the Earth. The two bodies will revolve rigidly, like a dumbbell with an invisible connecting rod. The Earth and the moon will at that time be separated, center to center, by a distance of 480,000 kilometers (300,000 miles).

The Approaching Moon

If there were no tidal effects on either the Earth or the moon, the dumbbell revolution would last forever. However, the sun's tidal effects would still exist. These effects would work in rather complicated fashion to speed up the rotations of the Earth and the moon and to draw the two bodies closer together, at a rate slower than that at which they are now separating. Apparently, this increasing closeness would continue indefinitely so that the moon would, it might be supposed, eventually fall to the Earth after all (though I began by saying it couldn't happen) since its angular momentum of revolution will finally be shifted all the way to angular momentum of rotation. It will not fall in the usual sense of the word, however, but will inch its way toward us in an excruciatingly slow and gradual decreasing spiral. Yet it will not truly fall even in this way, for no contact will be made.

As the two bodies approach closer and closer, the tidal effects will increase as the cube of the decreasing distance. By the time the Earth and moon are separated, center to center, by a distance of only some 15,500 kilometers (9,600 miles), so that the two surfaces are separated by only 7,400 kilometers (4,600 miles), the tidal effect of the moon on the Earth will be 15,000 times as strong as it is now. The tidal effect of the Earth on the moon will be 32.5 times stronger still, or nearly 500,000 times the moon's tidal effect on Earth today.

By that time, then, the tidal pull on the moon will be so great that the moon will simply be pulled apart and will break up into small fragments. The lunar fragments will, as a result of collisions (and further fragmenting), spread out through the moon's orbit and the Earth will end up with a ring, like that of Saturn, but far more brilliant and dense.

And what will happen to Earth, while all this is going on?

As the moon approaches the Earth, its tidal effect on Earth will grow enormously. The Earth will be in no danger of breaking up for the tidal effect upon it will be considerably smaller than the tidal effect on the moon. In addition, the Earth's greater gravitational field will more effectively hold it together against tidal pulls, than is the case for the moon. And, of course, once the moon breaks up and the gravitational field of its fragments is evenly spread around the Earth, the tidal effect becomes much smaller.

Nevertheless, just before the breakup of the moon, the tides on Earth will be so enormous that the ocean, raised into a bulge some kilometers high, will slosh completely over the continents, back and forth. Since the Earth's rotational period may be less than ten hours at the time, the tides will slosh back and forth every five hours.

It doesn't seem that either land or sea will be stable enough under such conditions to support anything but highly specialized forms of life, probably very simple in structure.

To be sure, we might imagine that human beings, if still in existence, might develop an underground civilization as the moon approached (it would be a very slow approach indeed, and it would not come by surprise). That would not save them, however, for under the tidal pulls, the groaning ball of Earth itself would be subject to constant earthquakes.

There is, however, no use worrying about the fate of the Earth as the moon approaches, for actually Earth would have become uninhabitable long before.

Suppose we go back to the vision of Earth and moon circling each other, dumbbell-fashion, every 47 days. If so, we can see that the Earth would already be a dead world. Imagine the surface of the Earth exposed to sunlight for a period of 47 days. The temperature would surely grow hot enough to boil water. Imagine the surface of the Earth exposed to darkness for a period of 47 days. The temperature would become Antarctic.

Of course, the regions of the poles are exposed to sunlight for even longer than 47 days at a time, but that is to a sun low on the horizon. In a slowly rotating Earth, the tropical regions would be subjected to a tropical sun for 47 days—quite different.

The temperature extremes would make Earth uninhabitable for most forms of life surely. At least it would be uninhabitable on the surface although we could imagine human beings establishing the underground civilization I mentioned earlier.

And yet we need not even be concerned about the dumbbell-rotation Earth-moon system for, oddly enough, it will never happen.

If Earth's day is gaining 1 second in length each 62,500 years, then in the 7 billion years that the sun will remain on the main sequence, the day would gain some 31 hours and be 2.3 present-days long. However, the moon will be receding in that interval and its tidal effect would be decreasing so that it would be fair to say that by the end of the 7-billion-year period, Earth's day would be somewhere in the neighborhood of twice its present length.

It would have no chance to grow longer, no chance to come even close to lengthening its day so greatly that it will revolve with the moon dumbbell-fashion, let alone begin spiraling together to develop those glorious rings. Long before anything like that happens, the sun will expand to a red giant and destroy Earth and moon alike.

It follows then that the Earth will remain habitable, as far as its rotation period is concerned, for as long as it exists, though with a double day the extremes of temperature during day and night would be greater than they are now, and somewhat uncomfortable.

However, humanity will undoubtedly have left the planet by then (assuming humanity survives those billions of years) and it will have been the swelling sun that will drive it away and not the slowing rotation.

Chapter 9

The Drift
of the Crust

Internal Heat

Since it doesn't seem as though sizable bodies from without (not even the moon) seriously threaten the Earth while the sun remains on the main sequence, let us abandon the rest of the universe for a while* and concentrate on the planet Earth.

Can any catastrophe take place that involves Earth itself in the absence of the intrusion of another body? For instance, can the planet blow up suddenly, and without warning? Or can it crack in two? Or can its integrity be impaired so drastically in any way as to constitute a catastrophe of the third class, putting an end to Earth as a habitable world?

After all, the Earth is an exceedingly hot body; it is only its surface that is cool.

* It will be necessary to return to it, now and then, in connection with nonsizable bodies.

The original source of the heat was the kinetic energy of motion of the small bodies that accumulated and crashed together to form the Earth some 4.6 billion years ago. The kinetic energy was converted into enough heat to melt the interior. Nor has the Earth's interior cooled down in the billions of years since. For one thing, the outer layers of rock are very good heat insulators and conduct heat only very slowly. For that reason, comparatively little heat leaks out of the Earth into surrounding space.

Of course some heat does, for there is no such thing as a perfect insulator, but even so, no cooling takes place. In the outer layers of the Earth there exist certain varieties of atoms that are radioactive. Four of them are particularly important: uranium-238, uranium-235, thorium-232 and potassium-40. These break down very slowly and in the course of all the billions of years of Earth's existence, some of each of these atom-varieties still exist unbroken down. To be sure, most of the uranium-235 and potassium-40 has gone by now, but only half of the uranium-238 has and only a fifth of the thorium-232.

The energy of breakdown is converted into heat, and although the amount of heat produced by a single atom breakdown is insignificant, the total heat produced by vast numbers of atoms breaking down at least matches the amount of heat lost from the Earth's interior. The Earth, therefore, is, if anything, slightly gaining heat, rather than losing it.

Is it possible, then, that the ravenously hot interior (and some estimates have the temperature as high as 2,700° C at the center) may produce an expansive force that will break through the cool crust like a vast planetary bomb, leaving only a belt of asteroids where there was once the Earth.

As a matter of fact, what seems to make the possibility a plausible one is the fact that there *is* an asteroid belt already existing between the orbits of Mars and Jupiter. Where did that belt come from? In 1802, the German astronomer Heinrich W. M. Olbers (1758–1840) discovered the second asteroid, Pallas, and at once speculated that the two asteroids, Ceres and Pallas, were small fragments of a large planet that had once orbited between Mars and Jupiter and had exploded. Now that we know there are tens of thousands of asteroids, most of them not more than a couple of kilometers across, that thought might seem to sound even more plausible today.

Another piece of evidence that seems to point in this direction is that of the meteorites that land on Earth (which are thought to arise from the asteroid belt); about 90 percent are stony and 10 percent are nickel-iron. That makes it seem that they are fragments of a planet with a nickel-iron core and a stony mantle around it.

Earth has that composition with the core making up about 17 percent of the volume of the planet. Mars is somewhat less dense than Earth and therefore must have a core (the densest part of the planet) that is smaller in proportion to the rest of the planet than is that of Earth. If the exploded planet were Marslike, that would explain the proportion of stony and nickel-iron meteorites.

There are even a couple of percent of the stony meteorites that are "carbonaceous chondrites" and contain significant quantities of the light elements—even water and organic compounds. These might be viewed as having originated in the outermost crust of the exploded planet.

And yet, neat as the theory of the explosive origin of the asteroids may sound, it is not accepted by astronomers. The best estimate we have of the total mass of the asteroids is that it comes to about 1/10 that of the moon. If all the asteroids were a single body, it would have a diameter of about 1,600 kilometers (1,000 miles). The smaller the body, the less heat in the center, and the less reason we can find for having it blow up. It seems extremely unlikely that a body only as large as a medium-sized satellite should explode.

It seems much more likely that as Jupiter grew, it was so efficient in sweeping up additional mass in its neighborhood (thanks to its already large mass) that it left very little in what is now the asteroid belt for accumulation into a planet. Indeed, it left so little that Mars could not grow to be as large as Earth and Venus. There was simply not enough matter available.

It could be, then, that the asteroid matter was too small in mass and generated too small a total gravitational field to collect into a single planet, particularly since the tidal effects of Jupiter's gravitational field worked against it. Instead, several moderately sized asteroids may have formed and collisions among them may have resulted in a vast powdering of smaller objects.

In short, the consensus now is that the asteroids are not the product of an exploding planet but the material of a planet that never formed.

Since there was no exploding planet in the space between Mars and Jupiter, we have less reason to think that any other planet will explode. What's more, we must not underestimate the power of gravity.

In an object the size of Earth, the gravitational field is dominant. The expansive influence of internal heat is far from enough to overcome the gravitational inpull.

We might wonder if radioactive breakdown of atoms in the Earth's body might not raise the temperature higher until a danger point is reached. As far as explosion is concerned, that is not a reasonable fear. If the temperature were to grow hot enough to melt the entire Earth, the present atmosphere and ocean might be lost, but the rest of the planet would keep on spinning as a huge drop of liquid still safely held together by its gravity. (The giant planet, Jupiter, is now thought to be just such a spinning drop of liquid with temperatures at the center as high as 54,000° C, though, to be sure, Jupiter's gravitational field is 318 times as intense as Earth's.)

Of course, if Earth grew hot enough to melt the entire planet, crust and all, that would be a true Catastrophe of the Third Class. We would not have to postulate an explosion.

This is not likely to happen either, though. The Earth's natural radioactivity is continuously declining. It is now, on the whole, less than half of what it was at the beginning of planetary history. If the Earth did not melt altogether in its first billion years of life, it is not going to melt now. And even if the temperature has been rising all through Earth's lifetime at a steadily diminishing rate and has not yet succeeded in melting the crust but is still laboring toward that goal, the temperature will rise so slowly that it will leave humanity plenty of time to escape the planet.

It is more likely that the internal heat of the Earth is, at best, holding its own, and as the radioactivity of the planet continues to decline, there may actually begin to be a very slow loss of heat. We may even envisage a far-distant future in which the Earth is essentially cold through and through.

Will this in any way affect life in such a way as to count as a catastrophe? As far as Earth's surface temperature is concerned, surely not. Almost all our surface heat comes from the sun. If the sun were to stop shining, the Earth's surface temperature would drop to far below Antarctic levels, and the internal heat of the planet would have an insignificant mitigating effect. If the internal heat of the Earth dropped to zero, on the other hand, and the sun kept shining, we would never know the difference as far as surface temperature was concerned. Nevertheless, the Earth's internal heat powers certain events with which human beings are familiar. Would the loss of this somehow prove catastrophic even if the sun were to remain shining?

This is not a question we need labor over. It will never come up. The decline of radioactivity and the loss of heat would proceed at so slow a pace that the Earth is sure to be an internally hot body, much as it is today, by the time the sun leaves the main sequence.

Catastrophism

Let us pass on to those catastrophes of the third class that would not compromise the integrity of the Earth as a whole, but would nevertheless leave it uninhabitable.

Bodies of myth frequently tell of worldwide disasters that put an end to all or almost all of life. It is very likely that these are born out of disasters of lesser scope that are exaggerated in memory and still further exaggerated in legend.

The earliest civilizations arose in river valleys, for instance, and river valleys are subject occasionally to disastrous floods. A particularly bad one that washed out all the area with which the inhabitants were acquainted (and the people of the early civilizations had only a limited appreciation of the extent of the Earth) would seem to represent worldwide destruction to them.

The ancient Sumerians, who dwelt in the Tigris-Euphrates valley in what is now Iraq, seem to have been subjected to a particularly bad flood about 2800 B.C. It made enough of an impression on them, and sufficiently disrupted their world, for them to date things afterward as "before the Flood" and "after the Flood."

Eventually, a Sumerian legend of the Flood grew up, one which is contained in the world's first known epic, the tale of Gilgamesh, king of the Sumerian city of Uruk. In his adventures, he encounters Ut-Napishtim, whose family had alone survived the Flood in a large ship he had built.

The epic was a popular one and it spread outward beyond the limits of the Sumerian culture and of those that followed it in the Tigris-Euphrates valley. It reached the Hebrews and probably the Greeks, both of whom incorporated a Flood story into their myths of the beginnings of the Earth. The version best known to us of the West is, of course, the biblical story as given in Chapters 6 through 9 in the Book of Genesis. The tale of Noah and the ark is too well known to be worth the retelling here.

Through many centuries the events of the Bible were accepted by almost all Jews and Christians as the inspired word of God and, therefore, as the literal truth. It was confidently assumed that there was indeed, some time in the third millennium B.C., a worldwide flood that had destroyed virtually all land life.

This predisposed scientists to suppose that the various signs of change they detected in the Earth's crust were the result of the violent cataclysm of the planetary Flood. When it seemed that the Flood was insufficient to account for all the changes, it was tempting to assume that other catastrophes had taken place at periodic intervals. This belief is termed "catastrophism."

The proper interpretation of the fossil remnants of extinct species and the deduction of a process of evolution was delayed by the assumption of catastrophism. The Swiss naturalist Charles Bonnet (1720–93), for instance, held that the fossils were indeed remnants of extinct species that had once been alive, but believed that they had died in one or another of the planetary catastrophes that had overwhelmed the world periodically. Of these, Noah's Flood was but the latest. After each catastrophe, seeds and other remnants of the pre-catastrophic life developed into new and higher forms. It was as though the Earth were a slate and that life was a message that was constantly being erased and rewritten.

The notion was taken up by the French anatomist Baron Georges Cuvier (1769–1832) who decided that four catastrophes, the last being the Flood, would explain the fossils. However, as more and more fossils were discovered, it was found that more and more catastrophes were needed to clear out some and pave the way for others. In 1849, a pupil of Cuvier's, Alcide d'Orbigny (1802–57), decided that no less than twenty-seven catastrophes were required.

D'Orbigny was the last breath of catastrophism in the main body of science. Actually, as more and more fossils were discovered and as the history of past life was worked out in greater and greater detail, it became clear that there were no catastrophes of the Bonnet-Cuvier type.

Disasters there have been in the history of the Earth and life has been affected by them dramatically, as we shall see, but no catastrophe has taken place of such a kind as to end all life and force it to begin again. No matter where one draws a line and says, "Here is a catastrophe," one can always find large numbers of species that lived through that period without change and without being in any way affected.

Life is undoubtedly continuous, and at no time since it first came into being over three billion years ago is there any clear sign of an absolute interruption. At every moment in all that period the Earth seems to have been occupied by living things in rich profusion.

In 1859, only ten years after d'Orbigny's suggestion, the English naturalist Charles Robert Darwin (1809–82) published his book *On the Origin of Species by Means of Natural Selection*. This advanced what we usually refer to as the "theory of evolution," and it involved the slow change of species over the eons, *without* catastrophe and regeneration. It met with great opposition at first from those who were scandalized at the way in which it contradicted the statements of Genesis, but it won out.

Even today, vast numbers of people, wedded to a literal interpretation of the Bible and completely unaware of the scientific evidence, remain hostile, out of ignorance, to the concept of evolution. Nevertheless, there is no scientific doubt that evolution is a fact though there remains plenty of room for dispute as to the exact mechanisms by which it has been brought about.* Even so, the story of the Flood and the hunger on the part of many people for dramatic tales keeps the notion of catastrophism of one sort or another alive outside the boundaries of science.

The continuing attraction of the suggestions of Immanuel Velikovsky, for instance, is due, at least in part, to the catastrophism he preaches. There is something dramatic and exciting about the vision of Venus flying at us and stopping Earth's rotation. The fact that it is in defiance of all the laws of celestial mechanics is not something that would disturb the kind of person who is excited by such tales.

Velikovsky advanced his notions originally to explain the biblical legend of Joshua stopping the sun and the moon. Velikovsky is willing to admit that it is really the Earth that is rotating, so he suggests the rotation is stopped. If the rotation stopped suddenly, as the biblical tale would imply, everything on Earth would go hurtling off.

Even if the rotation stopped gradually, over a period of a day or so, as Velikovskian apologists now insist—in order to explain why everything remained in place—the rotational energy would still be converted into heat and the Earth's oceans would boil. If the Earth's oceans had boiled at the time of the Exodus, it is difficult to see how Earth could be so rich in sea life now.

Even if we ignore the boiling, what is the chance that after Earth's rotation has stopped, Venus would then so affect the Earth as to

---

* Those who would deny evolution frequently state that it is "just a theory," but the evidence is far too strong for that. We might as well say that Newton's law of gravitation is "just a theory."

restart the rotation in the same direction and with the same period—to the second—that had existed before?

Many astronomers are absolutely bewildered and frustrated by the hold such nonsensical views have on so many people, but they underestimate the appeal of catastrophism. They also underestimate the lack of sophistication of most people in things scientific—especially among people who are thoroughly educated in nonscientific subjects. Indeed the educated nonscientists are more easily taken in by pseudoscience than others are since the mere fact of education in, say, comparative literature is apt to give one a falsely inflated opinion of one's power of understanding in an alien field.

There are other examples of catastrophism that attract the unsophisticated. For instance, any claim that the Earth every once in a while suddenly flops over so that what was once polar becomes temperate or tropical, and vice versa, finds willing ears. In this way, one can explain why some Siberian mammoths seem to have frozen so suddenly. Somehow to suppose that mammoths did something as simple as stumble into an icy crevasse or a freezing marsh is insufficient. What's more, even if the Earth flopped over, a tropical area would not instantly freeze. Heat loss takes time. If a house's furnace suddenly goes off on a cold winter's day, there is a perceptible interval of time before the temperature inside the house drops to freezing.

Besides, for the Earth to flop over is completely unlikely. There is an equatorial bulge as a result of Earth's rotation, and this causes the Earth to behave like a giant gyroscope. The mechanical laws governing the motion of a gyroscope are perfectly well understood, and the amount of energy required to cause the Earth to flop over is enormous. There is no source for this energy, barring the intrusion of a planetary object from outside, and of this, despite Velikovsky, there has been no sign in the last four billion years, nor any likely possibility in the foreseeable future.

A watered-down suggestion is that it is not the Earth as a whole that flops over, but merely the thin crust. The crust, merely a few dozen miles thick, and with only 0.3 percent of the Earth's mass, rests on the Earth's mantle, a thick layer of rock which, while not hot enough to be molten, is nevertheless quite hot and which can, therefore, be imagined to be soft. Perhaps, every once in a while, the crust slides over the upper surface of the mantle, producing all the effects, as far as surface life is concerned, of a complete flopover, and with far less expenditure of energy. (This was first suggested in 1886 by a German writer Carl Löffelholz von Colberg.)

What would cause such a crustal slip? One suggestion is that the vast ice cap over Antarctica is not perfectly centered on the South Pole. As a result the Earth's rotation would set up an off-center vibration which would eventually shake the crust loose and set it skidding.

This is not at all likely. The mantle is by no means soft enough for the crust to go skidding over it. If it were, the equatorial bulge would nevertheless hold it in place. And, in any case, the off-center position of the Antarctic ice cap is not enough to produce the effect.

What's more, it just hasn't ever happened. The slipping crust would have to tear apart as it passed from polar regions toward equatorial regions and crumple together as it passed from equatorial to polar regions. The tearing and crumpling of the crust in the case of such slippage would surely leave plenty of signs—except that it would probably destroy life and leave nothing to observe the signs.

In fact, we can generalize. There has been no catastrophe involving our planet in the last 4 billion years that has been drastic enough to interfere with the development of life, and the chances of there being one in the future that arises solely out of the mechanics of the planet itself is in the highest degree unlikely.

The Moving Continents

Having come to the conclusion of "no catastrophes," can we therefore decide that the Earth is perfectly stable and unchanging? No, indeed. There are changes and some of the changes are even of the kind I have already ruled out. How is that possible?

Consider the nature of catastrophe. Something that is catastrophic if it happens quickly may not be catastrophic at all if it happens slowly. If you were to descend from the top of a skyscraper very quickly by jumping off the roof, that would be a personal catastrophe for you. If, on the other hand, you descended quite slowly by elevator, that would present no problem at all. The same thing would have happened in either case; a change in position from top to bottom. Whether that change in position would be catastrophic or not would depend entirely on the rate of change.

Similarly, the speeding bullet that emerges from the muzzle of a gun and strikes your head will surely kill you; but the same bullet, moving only with the speed imparted to it by a person's throwing arm, will only give you a headache.

What I have eliminated as inadmissible catastrophes are therefore changes that happen *rapidly*. Those same changes, happening *very slowly*, are another matter altogether. Very slow changes can and do happen, and they need not be and, in fact, are not catastrophic.

For instance, having eliminated the possibility of catastrophic crustal slippage, we must admit that very slow crustal slippage is a possibility. Consider that some 600 million years ago, there seems to have been a period of glaciation (judging from scrapings on rocks of known age) that took place simultaneously in equatorial Brazil, in South Africa, in India, and in western and southeastern Australia. Those areas must have been covered by ice caps as Greenland and Antarctica are now.

But how can that be? If Earth's land-sea distribution were exactly the same then as it is now and if the poles were exactly in the same place, then to have tropical areas under an ice cap would mean that the whole Earth would have had to be frozen and that is very unlikely. After all, there is no sign of glaciation in other continental areas at that time.

If we suppose that the poles have shifted position so that what is now tropical was once polar and vice versa, then it is impossible to find a position for the poles that will account for all those primordial ice caps at the same time. If the poles have stayed in place, but the crust of the Earth has slipped as a whole, the problem is the same. There is no position in which all the ice caps are accounted for.

The only thing that can have happened to account for this long-ago glaciation is that the landmasses themselves have changed position relative to each other and that those various glaciated places were once close to each other and were all at one pole or the other (or some parts were at one pole and the rest at the other). Is that possible?

If we look at the map of the world, it isn't difficult to see that the eastern coast of South America and the western coast of Africa are amazingly similar. If you were to cut out both continents (assuming that the shape is not too badly distorted by being mapped on a flat surface) you can fit them together surprisingly well. This was noticed as soon as the shape of these coasts came to be known in sufficient detail. The English scholar Francis Bacon (1561–1626) pointed it out as long ago as 1620. Could it be that Africa and South America were once joined; that they split apart along the line of the present coasts and then drifted apart?

The first person to deal thoroughly with this notion of "continental drift" was a German geologist, Alfred Lothar Wegener (1880–1930), who published a book on the subject, *The Origin of Continents and Oceans,* in 1912.

The continents are made up of less dense rock than the ocean floor. The continents are chiefly granite, the ocean floor chiefly basalt. Might not these granite continental blocks very slowly drift about on the underlying basalt? It was something like the notion of the slipping crust, but instead of the entire crust slipping, it would be only the continental blocks that would do so—and very slowly.

If the continental blocks moved independently, there would be no serious problem with the equatorial bulge, and if they moved very slowly, not very much energy would be required and no catastrophe would result. Furthermore, if the continental blocks moved independently, this would account for a very ancient glaciation in widely spaced regions of the world, some near the equator. All those regions would have been together at one time, and at the poles.

Such continental drift might also be the answer to a biological puzzle. There are similar species of plants and animals that exist in widely separated portions of the world; portions separated by oceans those plants and animals could surely not have crossed. In 1880, the Austrian geologist Edward Seuss had explained this by supposing there had once been land bridges connecting the continents. For instance, he imagined a large supercontinent stretching around the entire southern hemisphere to explain how these species reached various land masses that are now widely separated. In other words, one had to imagine land rising and falling in the course of Earth's history, the same area being a high continent at one time and a deep ocean bottom at another.

The notion was popular but the more geologists learned about the sea bottom, the less likely it seemed that sea bottoms could ever have formed parts of continents. It would make more sense to suppose sideways movement, with a single continent breaking into fragments. Each of the fragments would carry particular groups of species and, in the end, similar species would be separated by wide oceans.

Wegener suggested that at one time all the continents existed as a single vast block of land set in one vast ocean. This supercontinent he called "Pangaea" (from Greek words meaning "all Earth"). For some reason, Pangaea broke up into several fragments which drifted apart until we ended with the continental arrangement of today.

Wegener's book aroused considerable interest, but it was difficult for geologists to take it seriously. The underlying layers of the Earth's continents were simply too stiff to allow those continents to drift. South America and Africa were each firmly fixed in place and neither could possibly drift through the basalt. For forty years, therefore, Wegener's theories were dismissed.

Nevertheless, the more the continents were studied, the more it seemed they must once have fitted together, all of them, especially if one considered the edge of the continental shelves as the true continental boundaries. It was too much to be dismissed as coincidence.

Suppose, then, we assume that Pangaea did exist and did split up and the fragments somehow did split apart. In that case, the floor of the oceans that formed between the fragments would have to be relatively young. Fossils from some rocks on the continents were as old as 600 million years, but fossils from the Atlantic sea bottom, which would have been formed only after Pangaea broke up, couldn't possibly be that old. As a matter of fact, no fossils older than 135 million years have ever been located from rocks at the Atlantic sea bottom.

More and more evidence accumulated in favor of continental drift. What was needed, however, was a suggestion as to the mechanism that would make it possible. It had to be something other than Wegener's suggestion of granite plowing through basalt; *that* was clearly *not* possible.

The key came with the study of the Atlantic sea bottom, which is, of course, hidden from us by an opaque sheet of water that is miles deep. The first hint that there might be something interesting down there came in 1853 when it proved necessary to make soundings so that the Atlantic cable could be laid, and Europe and America be connected by electric signals. At that time, it was reported that there seemed to be signs of an undersea plateau in the middle of the ocean. The Atlantic Ocean seemed definitely shallower in the middle than on either side, and the central shallow was named "Telegraph Plateau" in honor of the cable.

In those days, sounding took place by dropping a long, weighted line overboard. This was tedious, difficult, and uncertain, and few such soundings could be made, so that only the sketchiest details could be learned of the configuration of the sea bottom.

During World War I, however, methods for telling distance by means of ultrasonic echoes from objects underwater (now called "sonar") were worked out by the French physicist Paul Langevin (1872–1946). In the 1920s, a German oceanographic vessel began to make soundings in the Atlantic Ocean by sonar, and by 1925 it was shown that a vast undersea mountain range wound down the center of the Atlantic Ocean through all its length. Eventually this was shown to exist in the other oceans as well and, indeed, to encircle the globe in a long, winding "Mid-Oceanic Ridge."

After World War II, the American geologists William Maurice Ewing (1906–74) and Bruce Charles Heezen (1924–77) tackled the matter and by 1953 were able to show that running down the length of the ridge, right down its long axis, was a deep canyon. This was eventually found to exist in all portions of the Mid-Oceanic Ridge, so that it is sometimes called the "Great Global Rift."

The Great Global Rift seems to divide the Earth's crust into large plates which are, in some cases, thousands of kilometers across, and which seem to be 70 to 150 kilometers (45 to 95 miles) deep. These are called "tectonic plates" from the Greek word for "carpenter" since the various plates appear to be so neatly joined together. The study of the evolution of the Earth's crust in terms of these plates is referred to by these words in reverse as "plate tectonics."

The discovery of the tectonic plates established continental drift, but not in the Wegener fashion. The continents were not floating and drifting on the basalt. A particular continent, together with portions of adjacent sea bottom, was an integral part of a particular plate. The continents could only move if the plates moved, and it was clear that the plates moved. But how could they move if they were tightly joined?

They could be pushed apart. In 1960, the American geologist Harry Hammond Hess (1906–69) presented evidence in favor of "sea-floor spreading." Hot molten rock slowly welled up from great depths into the Great Global Rift in the mid-Atlantic, for instance, and solidified at or near the surface. This upwelling of solidifying rock forced the two plates on either side apart, in some places at the rate of from 2 to 18 centimeters (1 to 7 inches) a year. As the plates moved apart, South America and Africa, for instance, were forced apart. In other words, the continents didn't drift, they were pushed.

What produced the energy to account for this? Scientists are not certain but a likely explanation is that there are very slow swirls in the mantle underlying the crust, a mantle hot enough to be plastic under its great pressures. If one swirl moves up, west and down, and a neighboring swirl moves up, east and down, the opposite motions under the crust tend to push two neighboring plates apart, with hot material welling up between.

Naturally, if two plates are pushed apart, the other ends of those plates must be pushed into neighboring plates. When two plates are pushed together slowly, there is a crumpling and mountain ranges are formed. If they are pushed together more quickly, one plate slides under the other, moves into hot regions and melts. The ocean bottom is pulled down to form "deeps."

The whole history of the Earth can be worked out in plate tectonics, a study which has suddenly become the central dogma of geology, as evolution is the central dogma of biology, and atomism is the central dogma of chemistry. With tectonic plates moving apart here and coming together there, mountains rise, deeps depress, oceans widen, continents separate and rejoin.

Every once in a while the continents join into one huge land mass and then split up again, over and over. The last occasion on which Pangaea seems to have formed was 225 million years ago when the dinosaurs were just beginning to evolve; and it then began to break up about 180 million years ago.

Volcanoes

It might seem that the movement of the tectonic plates is not likely to be a catastrophic phenomenon since it is so slow. Throughout historic times, the shifting of the continents would not have been perceptible to any but the most careful scientific measurements. However, the movement of the plates produces occasional effects other than changes in the map, effects that are sudden and locally disastrous.

The lines along which the plates meet represent the equivalent of cracks in the Earth's crust and are called "faults." These faults are not simple lines, but have all kinds of branches and ramifications. The faults are weak points through which heat and molten rock well below the crust can, in some places, work their way upward. The heat may evidence itself rather benignly by warming ground water and producing steam vents or hot springs. Sometimes water is heated until the pressure reaches a critical point, whereupon a mass of it erupts high in the air. The situation then quiets as the underground supply refills and rewarms for the next eruption. This is a geyser.

In some areas the effect of the heat is more drastic. Molten rock wells up and hardens. More molten rock wells up through the mound of solidified rock, adding further to its height. Eventually, a mountain is built up with a central passageway through which the molten rock or "lava" can rise and subside; where it may solidify for longer or shorter periods, then melt again.

This is a "volcano," which may be active or inactive. Sometimes a particular volcano is more or less active over long periods of time and, as in any chronic ailment, is not then very dangerous. Occasionally, when underground events for some reason increase the level of activity, the lava rises and overflows. Then rivers of red-hot lava roll stickily down the sides of the volcano and, sometimes, make their way toward populated places, which must be evacuated.

Much more dangerous are those volcanoes that for periods of time are inactive. The central core through which lava had in the past risen then solidifies entirely. If there were no further activity underneath at all and forever, then all would be well. It happens sometimes, though, that conditions underground eventually, after a long lapse of time, begin to produce an excess of heat. The lava forming below is then penned in by the solidified lava above. The pressure builds up and eventually the top of the volcano is broken through forcibly. There is then a very violent and, what is worse, more or less unexpected hurling upward of gas, steam, solid rocks, and glowing lava. In fact, if water has been trapped under the volcano and has been turned to steam under enormous pressure, the entire top of the volcano may blow off to produce an explosion far greater than human beings have been able to manage, even in these days of fusion bombs.

Worse yet, an inactive volcano may seem completely harmless. It may have given no hint of activity in the memory of human beings, and the soil, being comparatively freshly brought up from depths, is usually very fertile. It therefore attracts human habitation and when the eruption comes (if it does) the results can be the more deadly.

There are 455 known active volcanoes in the world that erupt into the atmosphere. There are perhaps 80 more that are undersea. About 62 percent of the active volcanoes are to be found about the rim of the Pacific Ocean, three-fourths of them being on the ocean's western shores along the island chains that rim the Pacific coast of Asia.

This is sometimes called "the Ring of Fire" and it has been suggested that it was the still-raw scar marking the portion of the Earth that in primordial times had broken away to form the moon. This is no longer accepted as a reasonable possibility by scientists and the Ring of Fire merely marks the boundary of the Pacific plate with the other plates, east and west. Another 17 percent of the volcanoes occur along the island arm of Indonesia which marks the boundary of the Eurasian plate and the Australian plate. A further 7 percent are along an east-west line across the Mediterranean, marking the boundary between the Eurasian plate and the African plate.

The best-known volcanic eruption in western history was that of Vesuvius in A.D. 79. Vesuvius is a volcano about 1.28 kilometers (0.8 miles) high, and is located about 15 kilometers (10 miles) east of Naples. In ancient times it was not known to be a volcano for it had been inactive throughout the memory of human beings.

Then on August 24, A.D. 79, up it went. The flowing of lava, and the clouds of smoke, steam, and noxious vapors totally destroyed the cities of Pompeii and Herculaneum on its southern slopes. Because it happened at the height of the Roman Empire, because it was written up dramatically by Pliny the Younger (whose uncle, Pliny the Elder, died in the eruption while attempting to view the disaster at close hand) and because the excavations of the buried cities, which began in 1709, revealed a Roman suburban community that had been held, as it were, in suspended animation, this incident is *the* epitome of volcanic eruptions. It was, however, small potatoes as far as destruction was concerned.

The island of Iceland, for instance, is particularly volcanic, lying, as it does, on the Mid-Oceanic Ridge on the boundary between the North American plate and the Eurasian plate. It is indeed being pulled apart as the Atlantic sea-floor continues to spread.*

In 1783, the volcano, Laki, in south-central Iceland, 190 kilometers (120 miles) east of Reykjavik, the Icelandic capital, began to erupt. Over the space of two years lava covered an area of 580 square kilometers (220 square miles). The direct damage of the lava was small but volcanic ash was spread far and wide, even reaching Scotland, 800 kilometers (500 miles) to the southeast, and doing so in concentration sufficient to ruin croplands that year.

In Iceland itself, the ash and fumes killed three-quarters of all domestic animals and made at least temporarily useless what little agricultural land there was on the island. As a result, 10,000 people, one-fifth the island's population, died of starvation or disease.

Even worse can happen in more concentrated centers of population. Consider the volcano, Tambora, on the Indonesian island of Sumbawa, which lies just east of Java. In 1815, Tambora was 4 kilometers (2.5 miles) high. On April 7 of that year, however, the pent-up lava broke through and blew off the top kilometer of the volcano.

* The word "geyser" is the Icelandic contribution to the English language.

Perhaps 150 cubic kilometers (36.5 cubic miles) of matter were discharged in that eruption, making it the greatest mass of matter thrown into the atmosphere in modern times.* The direct rain of rock and ashes killed 12,000 people, and the destruction of farmland and domestic animals led to the death by starvation of 80,000 more on Sumbawa and on the neighboring island of Lombok.

In the western hemisphere, the most horrifying eruption in historic times came on May 8, 1902. Mount Pelée, on the northwestern end of the Caribbean island of Martinique had been known to emit minor hiccups now and then, but on that day it went up in a gigantic explosion. A river of lava and a cloud of hot gas poured down the volcano slopes at great speed, overwhelming the town of St. Pierre and utterly destroying its population. Altogether some 38,000 people died. (One man in the town, held in an underground prison, barely survived.)

The greatest explosion of modern times occurred, however, on the island of Krakatoa. It was not a large island, having an area of 45 square kilometers (18 square miles), just a bit smaller than Manhattan. It is located in the Sunda Strait between Sumatra and Java, 840 kilometers (520 miles) west of Tamboro.

Krakatoa did not seem particularly dangerous. There had been an eruption in 1680 but it hadn't amounted to much. On May 20, 1883, there was considerable activity, but it died away again without having done much damage, and thereafter continued in a sort of low-grade rumble. Then, at 10 A.M., on August 27, there was a tremendous explosion that virtually destroyed the island. Only about 21 cubic kilometers (5 cubic miles) of matter were hurled into the air, far less than the possibly exaggerated figure for the Tamboro eruption sixty-eight years before, but what was hurled upward was done so with far greater force.

Ash fell over an area of 800,000 square kilometers (300,000 square miles) and darkened the surrounding region for two-and-a-half days. Dust reached the stratosphere and spread over the entire Earth, giving rise to spectacular sunsets for a couple of years. The noise of the explosion was heard for distances of thousands of miles, over an estimated 1/13 of the globe, and the force of the explosion was about twenty-six times that of the largest H-bomb ever detonated.

The explosion set off a *tsunami* (a so-called "tidal wave") that washed over the neighboring islands and which made itself felt less catastrophically over all the ocean. All life of any kind on Krakatoa

* This may be an overestimate. It is possible that the top kilometer was not blown off entirely, but that much of it collapsed into the inner hollow formed by the erupting lava.

was destroyed and the *tsunami*, as it funneled into harbors where it reached heights of as much as 36 meters (120 feet), destroyed 163 villages and killed nearly 40,000 people.

Krakatoa has been called the loudest bang ever heard on Earth in historic times, but, as it turned out, that is wrong. There was a louder one.

In the southern Aegean Sea is the island of Thira, about 230 kilometers (140 miles) southeast of Athens. It is crescent-shaped, the open end to the west. Between the two horns are two small islands. The whole seems to be the circle of a large volcanic crater and so it is. The island of Thira is volcanic and undergoes numerous eruptions, but recent excavations show that in about 1470 B.C., the island was considerably larger than it now is and was the site of a flourishing branch of the Minoan civilization which had its center on the island of Crete, 105 kilometers (70 miles) south of Thira.

In that year, however, Thira blew up as Krakatoa was to do thirty-three centuries later, but with a force five times as great. Again, everything on Thira was destroyed, but the *tsunami* that was set up (reaching heights of 50 meters, or 165 feet, in some harbors) slammed into Crete and wreaked such havoc that the Minoan civilization was destroyed.* It was to be nearly a thousand years before the developing Greek civilization brought the culture of the area to the level it had achieved before the explosion.

Undoubtedly, the Thira explosion did not kill as many people as either the Krakatoa or Tamboro explosions did, because the Earth was far more thinly populated in those days. The Thira explosion has the sad distinction, however, of being the only volcanic eruption to destroy not a town or a group of towns, but an entire civilization.

The Thira explosion has another, quite romantic distinction. The Egyptians kept the records of that explosion, possibly in jumbled form,† and a thousand years later, the Greeks learned of it from them, possibly distorting it still further in the process. The tales surface in two of Plato's dialogues.

Plato (427–347 B.C.) did not try to be very historic about it, since he was using the tale in order to moralize. Apparently, he could not believe that the great city the Egyptians talked about existed in the

---

* Historians have known the Minoan civilization came to an end at that time, but until the Thira excavations had not known why.

† Velikovsky's gathered legends concerning the disasters of this period—in which he places the Exodus—if they have any meaning at all, could much more easily be attributed to the chaos and devastation that followed the Thira explosion, than to an impossible invasion of the planet Venus.

Aegean Sea where there were only small islands of no moment. He therefore placed it in the far west in the Atlantic Ocean, and called the destroyed city Atlantis. As a result, many people ever since have imagined that the Atlantic Ocean was the site of a drowned continent. The discovery of Telegraph Plateau seemed to lend credence to this, but of course the working-out of the Mid-Oceanic Ridge killed that idea.

What's more, the suggestion by Seuss of oceanic land bridges and of the rise and fall of vast areas of land stimulated the "lost continent" devotees even further. Not only was Atlantis imagined to exist, but also similar drowned continents in the Pacific and Indian oceans, called Lemuria and Mu. To be sure, Seuss was wrong and, in any case, he talked of events hundreds of millions of years ago, whereas the enthusiasts thought the ocean bottom was bouncing up and down a matter of tens of thousands of years ago.

Plate tectonics has put an end to all that. There are no drowned continents in any ocean—though the lost-continent devotees will continue to believe in their nonsense anyway, we may be sure.

Until quite recently, scientists (myself included) suspected it might be possible that Plato's account was entirely fictional, made up to point a moral. In that, we were wrong. Some of Plato's descriptions of Atlantis fit the Thira excavations, so the tale must have been based on the legitimate destruction of a city by an overnight catastrophe— but only a city on a small island, not a continent.

However, bad as volcanoes can be at their worst, there is another effect of plate tectonics which can be even more disastrous.

Earthquakes

When the tectonic plates pull apart or push together this is not necessarily done smoothly. In fact, one might expect a certain frictional resistance.

Two plates, we might imagine, are held tightly together by enormous pressures. The line is uneven, miles deep, and the lips of the plates are made of rough rock. The movement of the plates tends to push one north, let us say, while the other is stationary or pushing south. Or it may be that one plate is lifting while the other is stationary or sinking.

The enormous friction of the edges of the plates keeps them from moving, at least for a while, but the force tending to move the plates increases as the slow circulation in the mantle pulls plates apart in some places. The upwelling of molten rock and the sea-floor spreading exerts a steady push—push—push of one plate against another in other places. It may take years, but sooner or later friction is overcome and the plates move grindingly past each other, perhaps for just centimeters (or inches) or perhaps for meters (or yards). The pressure is then relieved and the plates settle down for another uncertain period of time before the next sizable movement.

When the movement of the plates does take place, the Earth vibrates and we have an "earthquake." Over the course of a century, two plates move against each other frequently, a short distance at a time, and the quakes may not be very powerful. Or, the plates may be held so tightly together that for a century nothing at all happens, then, suddenly, they let go and move one century's worth all at one time and there is a giant quake. As usual, the extent of damage depends on rate of change with time. The same energy release spread out over a century may do almost no harm, whereas squeezed into one short interval it can be cataclysmic.

Since earthquakes, like volcanoes, occur along faults—the places where two plates meet—the same regions that are given to volcanoes are also likely to experience earthquakes. Of the two phenomena, however, earthquakes are more deadly. Eruptions of lava occur in well-defined places—in the huge and easily recognized volcanoes. Usually, the disaster is confined to a small area, and only rarely are *tsunamis* and large blankets of ash involved. Earthquakes, on the other hand, can be centered anywhere along the line of a fault that may be hundreds of miles long.

Volcanoes usually give some warning. Even when a volcano blows its top suddenly, there are preliminary rumbles or emissions of smoke and ash. In the case of Krakatoa, for instance, there had been signs of activity for three months before the sudden explosion. Earthquakes, on the other hand, usually occur with only the most subtle warnings.

Whereas volcanic eruptions are almost always localized and almost always spread out over enough time to allow people to escape, an earthquake is usually over in five minutes and in those five minutes can affect a wide area. The earth tremors are not in themselves dangerous (though they may be terribly frightening), but they tend to knock down houses so that people die in the ruins. In modern times, they may crack dams and cause floods, break power lines and start fires, and, in short, do untold property damage.

The best-known earthquake in our modern Western history took place on November 1, 1755. The center was just off the coast of Portugal, and it was surely one of the three or four most powerful quakes recorded. Lisbon, Portugal's capital, got the full brunt of the quake and every house in the lower part of the city tumbled down. Then a *tsunami* set up by the undersea portion of the quake swept into the harbor and completed the wreck. Sixty thousand people were killed, and the city was flattened as though it had been struck by a hydrogen bomb.

The shock was felt over an area of 3.5 million square kilometers (1.5 million square miles), doing substantial damage in Morocco, as well as in Portugal. Because it was All Souls' Day, people were in church and all over southern Europe those in the cathedrals saw the chandeliers dance and sway.

The most famous earthquake in American history took place in San Francisco. That city lies on the boundary between the Pacific plate and the North American plate. This boundary runs the length of western California and is called the San Andreas fault. All along the fault and its branches, quakes are felt quite frequently, usually fairly mild ones, but sometimes stretches of the fault lock in place and when they are forced free after many decades, the results are devastating.

At 5:13 A.M. on April 18, 1906, the fault gave way at San Francisco and the buildings came down. A fire started that burned for three days until a rainfall brought it under control. Four square miles of the center of the city were wiped clean. About seven hundred people were killed and a quarter of a million people were left homeless. Property damage was estimated at half a billion dollars.

It was as a result of the study of this earthquake by the American geologist Harry Fielding Reid (1859–1944) that it was discovered there had been slippage along a fault. The ground had moved along one edge of the San Andreas fault relative to the other by up to 6 meters (20 feet). That gave rise to the modern understanding of earthquakes, although it was not until the development of plate tectonics a half-century later that the driving force behind earthquakes was understood.

The fame of the San Francisco earthquake must not be allowed to obscure the fact that the city was small at the time and that the deaths were relatively few. There have been much greater earthquakes in the western hemisphere as measured by death toll.

In 1970, in the resort town of Yungay, Peru, 320 kilometers (200 miles) north of the capital city of Lima, an earthquake released water that had been accumulating behind a wall of earth. A flood poured down and destroyed 70,000 lives.

Greater damage is done at the other end of the Pacific plate in the Far East where population is very dense and where the housing tends to be so flimsy that it comes down at once at the first tremor of a large quake. On September 1, 1923, an earthquake of giant size was centered just southwest of the Tokyo-Yokohama metropolitan area in Japan. Tokyo in 1923 was a far larger city than San Francisco was in 1906; there were about two million people living in the Tokyo-Yokohama area.

The earthquake occurred just before noon and 575,000 buildings were at once destroyed. The death toll in the quake, and in the fire that followed, may have reached over 140,000, and the cost of the property damage may have reached nearly three billion dollars (in terms of what dollars were worth at that time). This was very likely the most expensively damaging earthquake that has yet taken place.

Yet even that was not the worst quake from the standpoint of death toll. On January 23, 1556, in the province of Shensi in central China, a quake is reported to have killed 830,000 people. Of course, we can only feel limited confidence in a report so old, but on July 28, 1976, a similar devastating earthquake took place south of Peking in China. The cities of Tientsin and Tangshan were leveled, and while China would not put out official records of the casualties, unofficial reports are that 655,000 were killed and 779,000 injured.

What, then, can we say about earthquakes and volcanoes in general? They are disasters certainly, but they are strictly local. In the billions of years since life began, volcanoes and earthquakes have never come close to being ultimate destroyers of life. They cannot even be viewed as destroyers of civilization. That the Thira explosion was a powerful factor in the fall of the Minoan civilization is undoubtedly true, but civilizations were small in those days. The Minoan civilization was confined to the island of Crete, together with some Aegean islands, and with some influence over parts of the Greek mainland.

Can we be sure that matters will remain so; that tectonic disturbances won't become catastrophic in the future even though they weren't so in the past? In 1976, for instance, there were some fifty death-causing earthquakes, and some of them, such as those in Guatemala and in China, were true monsters. Is the Earth now falling apart for some reason?

Not at all! Things only *seem* bad and, in point of fact, the year 1906 (the year of the San Francisco earthquake) saw more disastrous quakes than did 1976, but in 1906 people didn't worry about it as much. Why do they worry about it more now?

First, communications have improved enormously since World War II. It was not many years ago that vast areas of Asia, Africa, and even South America were out of touch with us. If an earthquake took place in a remote region, only the faintest word reached the American public. Now, every earthquake is described at once and in detail on the front pages. The results of the devastation can even be seen on television.

Second, our own interest has grown. We are no longer isolated and self-absorbed. It was not so long ago that even if we heard details of earthquakes on other continents, we shrugged it off. What happened in far parts of the world didn't matter. Now we have become used to discovering that incidents anywhere in the world have an effect on us, so we pay more attention and grow more anxious.

Third, the world population has grown. It has doubled in the last fifty years and now stands at four billion. An earthquake which killed 140,000 in Tokyo in 1923 would, if it were to be repeated, kill perhaps a million persons. Consider that the population of Los Angeles was 100,000 in 1900 and is 3 million now. An earthquake affecting Los Angeles is now quite likely to kill thirty times as many people as it would have in 1900. That would not mean that the earthquake was thirty times as powerful; merely that the number of people available for killing had multiplied thirtyfold.

For instance, the most powerful recorded earthquake in the history of the United States took place not in California but in, of all places, Missouri. The center of the quake was near New Madrid on the Mississippi River, near the southeastern corner of the state, and the quake was so powerful that the course of the Mississippi was changed. The date, however, was December 15, 1811, and at that time the area was extremely thinly populated. There was not one recorded fatality. The same quake in the same place today would undoubtedly kill hundreds. If it were a few hundred kilometers upriver, it would kill tens of thousands.

Finally, we must remember that what really kills in the case of earthquakes is the works of people. Falling buildings bury people;

broken dams drown people; fires started by broken cables burn people. The works of man have multiplied with the years and grown more elaborate and expensive. Not only does this raise the death toll, but it also enormously raises property damage.

The Tectonic Future

We might expect, then, as a matter of course, that with each decade, the total death and destruction due to earthquakes and to a lesser extent to volcanoes, will grow worse, even though the plates do nothing more than continue to shift as they have been doing for some billions of years. We may also expect that people, noting the greater death and destruction, and subjected to greater publicity about it all, will be sure that the situation is growing worse and that the Earth *is* shaking apart.

It isn't! Even if things do seem to grow worse, it is the human change in the world, not the tectonic change, that is responsible. Of course there are always those who are eager, for some reason, to predict the imminent end of the world. In earlier times, the prediction was usually inspired by this or that portion of the Bible and was often viewed as the consequence of human sin. Nowadays, some material aspect of the universe is seized upon as the cause.

In 1974, for instance, a book called *The Jupiter Effect* by John Gribbin and Stephen Plagemann was published—and I wrote the foreword to it because I thought it was an interesting book. Gribbin and Plagemann calculated the tidal effect on the sun by several of the planets, speculated on the effect of tidal influence on solar flares and, therefore, on the solar wind, and speculated further on the effect of the solar wind on Earth. In particular, they wondered if there might not be a small effect that added to the stresses on various faults. If, for instance, the San Andreas fault was about to slip and produce a dangerous earthquake, the effect of the solar wind might add the final feather that would push the fault over the edge. Gribbin and Plagemann pointed out that in 1982 the planets would be so positioned as to make their tidal effect on the sun greater than usual. In that case *if* the San Andreas fault was about to slip, 1982 might be the year for it.

The thing to remember about the book is that, first, it is highly speculative. Second, even if the chain of events took place—if the position of the planets produced an unusually large tidal effect on the sun which raised the number and intensity of the flares which intensified the solar wind which nudged the San Andreas fault—all that would happen would be an earthquake that would have happened the next year anyway, if it had not been nudged into happening that year. It might be a very powerful earthquake, but it would be no more powerful than it would have been without the nudge. It might do enormous damage, but not because of its power; only because of the fact that human beings have filled up California with people and structures since the last great earthquake of 1906.

Nevertheless the book has been misunderstood and now there seems to be a feverish fear that there will be a "planetary lineup" in 1982 and that this will induce, by some sort of astrological influence, vast disasters on the planet, the very least of which will result in California sliding into the sea.

Nonsense!

The notion of California sliding into the sea seems to interest the irrationalists for some reason. Partly it must be because they have some dim notion that there is a fault running down the western edge of California (which there is) and that there can be movement along that fault (which there can). However, the movement would only be a few meters at the most and the edges of the fault would remain together. After all the damage was done, California would remain firmly in one piece.

To be sure, it is conceivable that at some time in the future there will be spreading along the fault; that material will well upward and force the two lips of the fault apart, producing a depression, perhaps, into which the Pacific Ocean could pour. The western sliver of California would then move away from the rest of North America, producing a long peninsula something like Lower California now, or even perhaps a long island. It would take millions of years for this to happen, however, and the process would be accompanied by nothing worse than the occurrence of earthquakes and volcanoes of the kind that exist now in any case.

Yet the California-slides-into-the-sea line of thought continues. For instance, there is an asteroid, Icarus, discovered in 1948 by Baade, which has a most eccentric orbit. At one end of its orbit it passes through the asteroid zone. At the other end it approaches the sun even more closely than does the planet Mercury. In between, its orbit passes fairly close to Earth's orbit, so that it is an Earth-grazer.

When Icarus and Earth are at the proper points in their orbits, they would be only 6.4 million kilometers (4 million miles) apart. Even at that distance, which is nearly seventeen times the distance of the

Moon, the effect of Icarus on Earth is nil. Nevertheless, at the most recent near approach, the warnings of California sliding into the sea could be heard.

In actual fact, the dangers of volcanoes and earthquakes could decrease with time. If, as stated earlier, Earth eventually loses its central heat, the driving force of plate tectonics, and therefore of volcanoes and earthquakes will be gone. This will surely not take place to any significant extent, however, before it is red-giant time for the sun.

More important is the fact that human beings are already attempting to take measures to minimize the danger. It would help, for instance, to have warning. In the case of volcanoes, this is relatively easy. A cautious avoidance of those objects and a prudent eye to the quite obvious premonitory symptoms that precede almost all eruptions could do much to help avert damage and death. Earthquakes are less cooperative, but they give signs too. As one side of a fault gets to the point of slipping against the other, some minor changes take place in the ground prior to the actual shock and these must, in one way or another, be capable of being detected and measured.

Changes in rock as it begins to give, just before an earthquake, include a decrease in the electrical resistance, a humping upward of the ground, and an increase in the flow of water from below into interstices that are being opened by the gradual stretching of the rock. The increased flow of water can be indicated by an increase in radioactive gases, such as radon, in the air—gases that have, until then, been imprisoned in the rocks. There are also rises in the level of well water and an increase in its muddiness.

Oddly enough, one of the important signs of an imminent earthquake seems to be a general change in the behavior of animals. Normally placid horses rear and race, dogs howl, fish leap. Animals, like snakes and rats, that ordinarily remain hidden in holes suddenly surge into the open. Chimpanzees spend less time in the trees and more on the ground. We needn't assume from this that animals have the ability to foretell the future or possess strange senses we lack. They live in more intimate contact with the natural environment, and their precarious lives force them to pay more attention to almost imperceptible changes than we ever do. Tiny tremblings that precede the real shock would upset them; strange sounds arising from the scraping of the lips of the fault would do the same.

In China, where quakes are more common and damaging than in the United States, great efforts are being made to predict them. The population is mobilized to be sensitive to change. Strange actions of animals are reported, as are shifts in the level of well water, the occurrence of strange sounds from the ground, even the unexplained flaking of paint. In this way, the Chinese claim to have anticipated damaging earthquakes by a day or two and saved many lives—notably, they say, in the case of a quake in northeastern China on February 4, 1975. (On the other hand, they seem to have been caught by surprise by the monster quake of July 28, 1976.)

In the United States, too, attempts at earthquake prediction are becoming more serious. Our forte is high technology, and we can turn to it to detect the delicate changes in local magnetic, electrical, and gravitational fields, as well as day-to-day changes in the level and chemical content of well-water and in the properties of the air about us.

It will be necessary, however, to judge the place, time, and magnitude of an earthquake occurrence quite accurately, for a false alarm could be costly. Rapid evacuation could do more in the way of economic dislocation and personal discomfort than a minor earthquake would and popular reaction would be unfavorable if the evacuation proved to be unnecessary. At the time of the next warning, people would refuse to evacuate—and then the earthquake might hit.

To increase the chances of predicting an earthquake with reasonable certainty, a variety of measurements probably would have to be made and the relative importance of their changing values weighed. One can imagine the quivering readings of a dozen needles, each measuring a different property, being fed into a computer, which would constantly weigh all the effects and yield an overall figure which, upon passing a certain critical point, would signal evacuation.

Evacuation would minimize damage, but need we be satisfied with that? Might earthquakes be prevented altogether? There seems no practical way by which we could modify subterranean rock, but subterranean water is another matter. If deep wells are drilled several kilometers apart along the line of a fault, and if water is forced into them and then allowed to backflow, subterranean pressures might be relieved and an earthquake aborted. Indeed the water might do more than relieve pressures. It might lubricate the rocks and encourage slippage at more frequent intervals. A series of minor earthquakes that do no harm, even cumulatively, is far better than one major one.

Although it is easier with a few days' notice to predict a volcanic eruption than an earthquake, it would be harder and more dangerous to attempt to relieve volcanic pressures than earthquake pressures.

Still, it is not too much to imagine that inactive volcanoes might somehow be bored into in such a way as to keep open a central passageway through which the hot lava could rise without building pressures to the explosive point—or where new channels are cut nearer ground level in directions designed to cause least trouble for people.

To summarize, then, it seems reasonable to suppose that the Earth will remain sufficiently stable through the sun's stay on the main sequence and that life will not be endangered by any convulsion of the Earth itself or by any untoward movement of its crust. And as for the local disasters of volcanoes and earthquakes, it may even be possible to reduce those dangers.

Chapter 10

The Change
of Weather

The Seasons

Even if we assume an absolutely reliable sun and an absolutely stable
Earth, there are periodic changes about us that place a strain on our
ability, and the ability of living things in general, to remain alive.
Because the Earth is unevenly heated by the sun, thanks to its spherical shape, to its slightly changing distance from the sun as it moves
about its elliptical orbit, and to the fact that its axis is tipped, the
average temperatures in any particular spot on Earth rise and fall in
the course of the year, which is therefore marked off into seasons.

In the temperate zones we have a distinctly warm summer and a
distinctly cold winter, with heat waves in the former and snowdrifts
in the latter; and with the intermediate seasons of spring and fall in
between. The differences in seasons are less noticeable as we travel
toward the equator, at least as far as temperature is concerned. But

191

even in the tropic regions where temperature differences in the course of the year are not great and there is an eternal summer, there are likely to be rainy seasons and dry seasons.

The differences in seasons are more noticeable as we travel toward the poles. The winters grow more frigid with a lower sun, and the summers briefer and cooler until finally at the poles themselves there are the legendary days and nights each six months long with the sun skimming the horizon, just above or below it respectively.

Naturally, the seasons of the year do not vary smoothly in temperature, as we all know. There are extremes that sometimes reach disastrous intensities. There are periods, for instance, when rainfall is less than normal over extended periods and the result is a drought in which crops fail. Since population in agricultural areas tends to rise to the limit that can be supported in years of good harvests, a drought is followed by a famine.

In preindustrial times, when transportation over long distances was difficult, a famine in one province could proceed to extremes even though neighboring provinces had food surpluses. Even in modern times, millions starved now and then. In 1877 and 1878, 9.5 million people died in a famine in China and 5 million died in the Soviet Union after World War I.

Famines should be a smaller problem now, for it is possible, for instance, to ship American wheat to India rapidly in case of need. Nevertheless, there are still problems. Between 1968 and 1973, there was a drought in the Sahel, that portion of Africa lying south of the Sahara Desert, and a quarter of a million people famished and died, while millions more were brought to the edge of starvation.

In reverse, there are periods when rainfall is much heavier than normal and, at the worst, this can produce the quick destruction of river flooding. These are particularly destructive in the flat, crowded lands that border the Chinese rivers. The Hwang-Ho, or Yellow River (also called "China's sorrow") has in the past overflowed and killed hundreds of thousands. A flood of the Hwang-Ho in August of 1931 is supposed to have drowned 3.7 million people.

Sometimes it is not the flooding of the river that does the damage as much as the violent winds that can accompany a rainstorm. In hurricanes, cyclones, typhoons, and so on (different regions apply different names to a large area of rapidly circling winds) the combination of wind and water can be deadly.

Particular damage is done in the crowded, low-lying delta land of the Ganges River in Bangladesh, where, on November 13, 1970, perhaps as many as 1 million people died under the driving lash of a cyclone that drove the sea inland. At least four other such storms in the previous decade had each killed ten thousand or more people in Bangladesh.

Where wind is combined with snow in the lower temperatures of winter to form blizzards, the deadliness is less, if only because such storms are most common in polar and semipolar areas where the population is thin. Nevertheless, on March 11–14, 1888, a three-day blizzard in the northeastern United States killed 4,000 people, and a hail storm killed 246 people in Moradabad, India, on April 30 of that same year.

The most dramatic storm of all is the tornado, which consists of tightly spiraling winds of up to 480 kilometers (300 miles) per hour in velocity. They can literally destroy everything in their path, but have the saving grace of generally being small and short-lived. Even so, up to a thousand of them can occur in the United States in a single year, most in the central regions, and the death toll is not insignificant. In 1925, 689 people were killed by tornadoes in the United States.

These and all weather extremes, however, can qualify only as disasters and not as catastrophes. None come even close to threatening life, or even civilization, as a whole. Life has adjusted to these seasons. There are organisms adapted to the tropics, to the deserts, to the tundra, to the rain forests, and life can survive all the extremes, though it may be battered a bit in the process.

Is it possible, though, that the seasons can change their nature and wipe out most or all of life by means, let us say, of a prolonged winter or a prolonged dry season? Can Earth become a planetary Sahara or a planetary Greenland? From our experience in historic times, the temptation is to say no.

There have been small swings of the pendulum. For instance, during the Maunder minimum in the seventeenth century, the average temperature was lower than normal—but not sufficiently to endanger life. We can have successions of dry summers or mild winters or stormy springs or sodden falls, but things always bounce back and nothing ever becomes truly unendurable.

About the closest the Earth came in recent centuries to experiencing a true climactic aberration was in 1816, the year after the tremendous volcanic explosion in Tamboro. So much dust was hurled high into the stratosphere that an unusual quantity of solar radiation was reflected back into space by the dust and prevented from reaching the Earth's surface. The effect was equivalent to that of making the sun dimmer and cooler, and as a result 1816 came to be known as "the year without a summer." In New England, it snowed at least once in every month of the year, including July and August.

Clearly if this had kept on year after year without letup, the results would eventually have been catastrophic, but the dust settled and the climate returned to its wonted round.

Suppose, though, we looked back to prehistoric times. Was there ever a period when the climate was markedly more extreme than it is today? When it was extreme enough to approach the catastrophic? Naturally, it could never have been sufficiently extreme to put an end to all life, for living things still populate the Earth in profusion—but could it have been extreme enough to cause such problems that, if it were to recur just a little worse, it would seriously threaten life?

The first hint that there was at least a possibility of this arose in the late eighteenth century, when modern geology was coming into being. Some aspects of the Earth's surface began to seem puzzling and paradoxical in the light of the new geology. Here and there one found boulders of a nature unlike the general rocky background. In other places, there were deposits of sand and gravel that didn't seem to fit. The natural explanation at that time was that these dislocations were brought about by Noah's Flood.

In many places, however, exposed rocks were covered with parallel scratches, ancient weathered scratches that might have been caused by the scraping of rock on rock. In that case, though, something would have had to hold two rocks together with great force and yet have the additional force to move one against the other. Water alone could not do that, but if not water, what?

In the 1820s, two Swiss geologists Johann H. Charpentier (1786–1855) and J. Venetz considered the matter. They were well acquainted with the Swiss Alps and they were aware that when glaciers melted and retreated somewhat in the summer, they left behind deposits of sand and gravel. Could it be that the sand and gravel had been carried down the mountain slopes and that the glacier accomplished this task because it moved like a very, very slow river? Could glaciers carry large boulders as well as sand and gravel? And if glaciers were once much larger than they are now, could they have scraped pebbles over other boulders, producing the scratches? Then if the glaciers carried sand, gravel, pebbles, boulders far beyond the limits to which those glaciers now stretched, might they then retreat leaving the matter behind and in surroundings in which they did not belong?

Charpentier and Venetz maintained that this was what happened. They suggested that the Alpine glaciers had been much larger and longer in times past and that the isolated boulders in northern Switzerland had been carried there by the enormous glaciers that had extended from the southern mountains in the past, and were left there when the glaciers retreated and dwindled.

The Charpentier-Venetz theory was not taken seriously at first since scientists generally doubted that the glaciers could flow like rivers. One of the doubters was a young friend of Charpentier, a Swiss naturalist Jean L. R. Agassiz (1807–73). Agassiz decided to test glaciers to see if, indeed, they flowed. In 1839, he hammered stakes 6 meters (20 feet) into the ice, and by the summer of 1841, he found that they had moved a substantial distance. What's more, those in the center of the glacier had moved considerably farther than those near the sides where the ice was held back by friction with the mountainside. What had been a straight line of stakes became a shallow U, with its opening pointed uphill. This showed that the ice did not move all in one piece. Instead, there was a kind of plastic flow as the weight of the ice above forced the ice below slowly to extrude, like toothpaste out of a tube.

Eventually, Agassiz traveled all over Europe and America looking for signs of glacier scrapings on rocks. He found boulders and detritus in odd places that marked the forward push and retreat of glaciers. He found depressions or "kettle holes" that seemed to have the characteristics one would expect if they had been dug out by glaciers. Some of them were filled with water, and the Great Lakes of North America are examples of particularly large water-filled kettle holes.

Agassiz's conclusion was that the time of extended glaciers in the Alps was also a time of vast sheets of ice in many places. There was an "Ice Age" when ice sheets like those that now cover Greenland covered large areas of North America and Eurasia as well.

Careful geological studies since then have shown that weather as it exists today is indeed far from typical of particular times in the past. Glaciers have expanded from the polar regions southward a number of times in the last million years and have then retreated only to advance again. Between the periods of glaciation there were "interglacial ages" and we are living in one now—but not completely. The huge ice cap in Greenland is a living reminder of the most recent period of glaciation.

Triggering the Glaciers

The ice ages of the last million years have obviously not brought an end to life on the planet. They didn't even put an end to human life. *Homo sapiens* and its hominid ancestors lived all through the ice ages of the last million years without any noticeable interruption of rapid evolution and development.

Nevertheless, it is fair to wonder whether another period of glaciation is ahead of us or whether it is all part of the past. Even if an ice age does not mean an end to life or even to humanity, and is not catastrophic in that sense, the thought of almost all of Canada and of the northern quarter of the United States under a mile-deep glacier (to say nothing of portions of Europe and Asia similarly iced) might seem quite bad enough.

To decide whether the glaciers might return, it would help to learn first what causes such periods of glaciation. And before trying to do that, we must understand that it doesn't take much to start the glaciers moving; we don't have to postulate large and impossible changes.

Right now, snow falls over much of northern North America and Eurasia every winter and leaves all that region covered with frozen water almost as though the Ice Age had returned. The snow cover, however, is only a few centimeters to a couple of meters thick, and in the course of the summer it all melts. There is, in general, a balance and on the average as much snow melts in the summer as falls in the winter. There is no overall change.

But suppose that something happens which cools the summers just a bit, perhaps only two or three degrees. This would not be enough to notice, and of course if would not be a steady change. There would continue to be warmer summers and cooler summers in some random distribution, but the warmer summers would occur less frequently and the cooler summers more frequently so that, on the average, the snow that fell in the winter would not quite all melt in the summer. There would be a net increase in snow cover from year to year. This would be a very slow increase and it would be noticeable in the northern polar and subpolar regions and in the higher mountainous regions. The accumulating snow would turn to ice and the glaciers that exist in the polar regions and at higher elevations even in southerly latitudes would extend farther in the winter and retreat less in the summer. They would grow from year to year.

The change would feed on itself. Ice reflects light more efficiently than bare rock or soil does. In fact, ice reflects some 90 percent of the light that falls on it, while bare soil reflects less than 10 percent. This means that as the ice cover expands, more sunlight is reflected and less is absorbed. The average temperature of the Earth would drop a little farther, summers would grow a trifle cooler still, and the ice cover would expand still more rapidly. As a result of a very small cooling trigger, then, the glaciers would grow and turn into ice sheets that would slowly advance, year by year, until finally they could cover vast stretches of ground.

Once an ice age had well established itself and the glaciers had reached far southward, however, a reverse trigger, very small in itself, could initiate a general retreat. If the average temperature of summer rose two or three degrees over an extended period of time, more snow would melt in the summer than would fall in the winter and the ice would recede somewhat from year to year. As it receded, the Earth as a whole would reflect somewhat less sunlight and absorb somewhat more. This would further warm the summers and the glacier retreat would be accelerated.

What we must do, then, is identify the trigger that sets off the glacial advance—and retreat. This is not hard to do. The trouble is, in fact, that there are too many possible triggers and the difficult task is to choose among them. For instance the trigger may lie in the sun itself. Earlier, I mentioned that the Maunder minimum came at a time when Earth's weather was generally on the chilly side. The time is actually referred to, sometimes, as "the Little Ice Age."

If there is a causal connection, if Maunder minima cool the Earth, then perhaps every hundred thousand years or so it may be that the sun goes through an extended Maunder minimum, one that doesn't last a few decades, but a few millennia. The Earth may then be chilly long enough to initiate and maintain an ice age. When the sun finally begins to spot again and experiences only short Maunder minima at most, the earth would warm up slightly and the glacial retreat begin.

There may be something to this, but we have no evidence. Perhaps further studies of the solar neutrinos and why they are so few in number may help us know enough of what is going on inside the sun to allow us to understand the intricacies of the sunspot cycle. We might then be able to match the sunspot variations with the periods of glaciation and be able to predict if and when another period will arrive.

Or, it might not be the sun itself, which might shine with beautiful steadiness. It might, instead, be the nature of the space between the Earth and sun.

I explained earlier that there was only an incredibly small chance of a close encounter with a star or any other small object from interstellar space, on the part of either the sun or the Earth. There are, however, occasional clouds of dust and gas between the stars here in the outskirts of our galaxy (and of other galaxies like it) and the sun in its orbit about the galactic center might easily pass through some of those clouds.

The clouds are not dense by ordinary standards. They would not poison our atmosphere or us. They would not, in themselves, be particularly noticeable to the average observer, let alone catastrophic. Indeed, NASA scientist, Dixon M. Butler, suggested in 1978 that our solar system has passed through at least a dozen quite extensive clouds in the course of its lifetime and, if anything, this may be an underestimate.

Almost all the materials in such clouds are hydrogen and helium, which would not affect us at all, one way or another. However, about 1 percent of the mass of such clouds consists of dust; grains of ice or rock. Each of these grains would reflect, or absorb and reradiate, sunlight, so that less sunlight than normal would make its way past the grains to fall on Earth's surface.

The grains might not blank out the light falling on Earth very much. The sun would look as bright and perhaps even the stars would look no different. Nevertheless, a particularly dense cloud might blank out just enough light to cool the summers the proper amount to trigger an ice age. Moving out of the cloud might serve as the trigger for glacial retreat.

It may be that for the last million years the solar system has been passing through a cloudy region of the Galaxy and that whenever we pass through a particularly dense cloud that will blank out just enough light, an ice age will start, and when we leave it behind us, the glaciers retreat. Prior to the last million-year period there was a 250-million-year period in which there were no ice ages, and perhaps during that time the solar system was passing through clear regions. Prior to that there was the Ice Age I mentioned as giving rise to the thought of Pangaea.

It may be that every 200 to 250 million years there are a series of ice ages. Since this is not very different from the period of revolution of the solar system about the galactic center, perhaps we are passing through the same cloudy region every revolution. If we have now passed through the region completely, then there may be no periods of glaciation for a quarter of a billion years. If not, another one—or a whole series of them—is due much sooner than that.

In 1978, for instance, a group of French astronomers presented evidence leading to the possibility of another interstellar cloud just ahead. The solar system may be approaching it at a velocity of 20 kilometers (12.5 miles) per second, and at that rate, it may reach the edges of the cloud in about 50,000 years.

But it may not be either the sun directly or the dust clouds of interstellar space that are the true trigger. It may be Earth itself, or rather its atmosphere, that offers the necessary mechanism. The sun's radiation has to pass through the atmosphere and that might affect it.

Consider that the sun's radiation reaching Earth does so chiefly in the form of visible light. The peak of the sun's radiation *is* in the wavelengths of visible light and this passes through the atmosphere easily. Other forms of radiation, such as ultraviolet and x-rays, which the sun produces in lesser profusion are blocked by the atmosphere.

In the absence of the sun—as at night—the Earth's surface radiates heat away into outer space. It does so chiefly in the form of long infrared waves. These pass through the atmosphere, too. Under ordinary conditions, these two effects balance and the Earth loses as much heat from its night-shrouded surface as it gains on its daylight-drenched surface, and its average surface temperature remains the same from year to year.

Nitrogen and oxygen, which make up virtually all the atmosphere, are easily transparent to both visible light and to infrared radiation. Carbon dioxide and water vapor, however, while transparent to visible light are not transparent to infrared. This was first pointed out in 1861 by the Irish physicist John Tyndall (1820–93). Carbon dioxide makes up only 0.03 percent of the Earth's atmosphere and the water vapor content is variable but low. Therefore, they don't block the infrared radiation altogether.

Nevertheless, they do block the infrared radiation somewhat. If the Earth's atmosphere lacked carbon dioxide and water vapor entirely, more infrared radiation would escape at night than does so now. The nights would be colder than they are now and the days, warming up from a colder start, would be cooler. The average temperature of the Earth would be distinctly less than it is now.

The carbon dioxide and water vapor in our atmosphere, even though present in small quantities, block enough of the infrared to act as appreciable conservers of heat. Their presence serves to produce a distinctly higher average temperature for the Earth than would otherwise be the case. This is called the "greenhouse effect," because the glass of a greenhouse works similarly, letting through the visible light of the sun and holding back the infrared reradiation from the interior.

Suppose, for some reason, the carbon dioxide content of the atmosphere goes up slightly. Let us suppose it doubles to 0.06 percent. This would not affect the breathability of the atmosphere and we would be unaware of the change in itself—only of its effects. An atmosphere with that slight increase in carbon dioxide would be still more opaque to infrared radiation. Since infrared radiation is held back, the temperature of the Earth would rise slightly. The slightly higher temperature would increase the evaporation of the oceans, raise the level of water vapor in the air, and that, too, would contribute to an increased greenhouse effect.

Suppose, on the other hand, the carbon dioxide content of the atmosphere goes down slightly, from 0.03 percent to 0.015 percent. Now the infrared radiation escapes more easily and the temperature of the Earth drops slightly. With lower temperatures, the water vapor content decreases, adding its bit to the reverse greenhouse effect. Such rises and falls in temperature could be enough to end or begin a period of glaciation.

But what could bring about such changes in carbon dioxide content of the atmosphere? Animal life produces carbon dioxide in great quantity, but plant life consumes it in equally great quantity, and the effect of life generally is to maintain the balance.* There are, however, natural processes on Earth that either produce or consume carbon dioxide independently of life and they may unbalance the equilibrium sufficiently to produce a trigger.

For instance, a great deal of atmospheric carbon dioxide can dissolve in the ocean, but carbon dioxide dissolved in the ocean can easily be given up to the atmosphere again. Carbon dioxide can also react with the oxides of the Earth's crust to form carbonates and there the carbon dioxide is more likely to stay put.

Of course, those portions of the Earth's crust that are exposed to air have already absorbed what carbon dioxide they can. During periods of mountain formation, however, new rock reaches the surface, new rock that has not been exposed to carbon dioxide, and this can

* This is not entirely true of that portion of life that includes human activity. I'll get back to that later.

act as a carbon dioxide-absorbing medium, reducing the percentage in the atmosphere.

On the other hand, volcanoes spew vast quantities of carbon dioxide into the atmosphere since the intense heat that melts rocks into lava breaks up the carbonates and liberates the carbon dioxide again. In periods of unusually high volcanic activity, the atmospheric content of carbon dioxide may go up.

Both volcanoes and mountain building are the result of the movements of the tectonic plates, as I have mentioned, but there are times when the conditions for vulcanism are more common than for those of mountain formation and there are times when the reverse is true.

It may be that when mountain formation is more characteristic of a period in Earth's history, the carbon dioxide content goes down, the Earth's surface temperature drops, and the glaciers begin to advance. When it is vulcanism that predominates, the carbon dioxide content goes up, the Earth surface temperature rises, and the glaciers, if present, begin to retreat.

But just to show that things are not as simple as they might sound, if volcanic eruption tends to be *too* violent, large quantities of dust may be hurled into the stratosphere and this may produce so many "years without a summer" like 1816, that *this* may trigger an ice age.

From the volcanic ash in ocean sediments, it would seem that vulcanism in the last 2 million years has been some four times as intense as in the preceding 18 million years. Perhaps it is a dusty stratosphere, then, that is subjecting the Earth to its periodic ice ages now.

Orbital Variations

So far, the possible triggers for glaciation and deglaciation that I have described don't lend themselves to very confident predictions of the future.

We don't really know, as yet, what the rules are that govern the small changes in the sun's radiation output. We are not quite sure what lies ahead of us as far as collisions with cosmic clouds are concerned. We certainly can't predict the effects of volcanic eruptions and mountain formation in the future. It would seem that whatever the trigger, human beings would have to live from year to year and millennium to millennium, scanning the weather reports and wondering.

There is, however, one suggestion that would make it seem that the coming and going of ice ages are as regular and as inevitable as the change of seasons in the course of a year.

In 1920, a Yugoslavian physicist Milutin Milankovich suggested that there was a great weather cycle as a result of small periodic changes involving the Earth's orbit and its axial tilt. He spoke of a "Great Winter" during which the ice ages took place and a "Great Summer" which represented the interglacial periods. In between, of course, would be a "Great Spring" and a "Great Fall."

At that time, Milankovich's theories received no more attention than Wegener's theories of continental drift did, but just the same there *are* changes in Earth's orbit. For instance, Earth's orbit is not exactly circular, but is slightly elliptical, with the sun at one of the foci of the ellipse. This means that the distance of the Earth from the sun varies slightly from day to day. There is a time when the Earth is at "perihelion" and is closest to the sun and a time, six months later, when it is at "aphelion" and is farthest from the sun.

The difference isn't much. The orbit is so slightly elliptical (it is an ellipse of such low eccentricity) that if it were drawn to scale you could not tell it from a circle by eye. Nevertheless, the small eccentricity of 0.01675 means that at perihelion, the Earth is 147 million kilometers (91,350,000 miles) from the sun and at aphelion it is 152 million kilometers (94,450,000 miles) from the sun. The difference in distance is 5 million kilometers (3.1 million miles).

This is a great deal by earthly standards, but the difference is only about 3.3 percent. The sun is slightly larger in appearance at perihelion than at aphelion, but not enough so for it to be noticed by anyone but astronomers. Also, the sun's gravitational pull is a little stronger at perihelion than at aphelion, so that the Earth moves faster in the perihelion half of the orbit than in the aphelion half, and the seasons are not of exactly equal lengths—and this, too, goes unnoticed by the ordinary person.

Finally, it means that at perihelion we get more radiation from the sun than we do at aphelion. The radiation we get varies inversely as the square of the distance, so that it turns out Earth gets almost 7 percent more radiation at perihelion than at aphelion. Earth reaches its perihelion on January 2 of each year and its aphelion on July 2. It so happens that January 2 is less than two weeks after the winter solstice, while July 2 is less than two weeks after the summer solstice.

This means that at the time Earth is at or near perihelion and getting more heat than usual, the northern hemisphere is deep in winter, while the southern hemisphere is deep in summer. The extra heat means that the northern winter is milder than it would be if Earth's orbit were circular, while the southern summer is hotter. At the time Earth is at or near aphelion and getting less heat than usual, the northern hemisphere is deep in summer, while the southern hemisphere is deep in winter. The heat deficiency means that the northern summer is cooler than it would be if the Earth's orbit were circular and the southern winter is colder.

We see, then, that Earth's orbital ellipticity gives the northern hemisphere, outside the tropics, a less extreme swing between summer and winter than is true of the southern hemisphere outside the tropics.

This may sound as though the northern hemisphere is not likely to have an ice age, while the southern hemisphere is, but that's wrong. Actually, it is the mild winter and cool summer—the less extreme swing—that predisposes a hemisphere to an ice age.

In the winter, after all, it snows as long as the temperature is below freezing, provided there is excess moisture in the air. Sending the temperature farther below freezing doesn't increase the snow. Instead, it is likely to decrease it since the colder the temperature, the less moisture the air can contain. Maximum snowfall comes in a winter that is as mild as it can be without actually rising above the freezing point too often.

The amount of snow melting in the summer depends, of course, on the temperature. The hotter the summer, the more snow is melted, and the cooler the summer the less snow is melted. It follows that when you have mild winters and cool summers, you have much snow and little melting and that is precisely what you need to start an ice age.

Yet there is no ice age in the northern hemisphere now even though we have mild winters and cool summers. Perhaps it may be that the swing is still too extreme, that there are other factors that will act to make the winters still milder and the summers still cooler. At the present moment, for instance, the Earth's axis is tipped from the vertical by about 23.5°. At summer solstice, June 21, the northern end of the axis is tipped in the direction of the sun. At the winter solstice, December 21, the northern end of the axis is tipped in the direction away from the sun.

The Earth's axis, however, doesn't stay tipped in the same direction forever. Because of the moon's pull on the Earth's equatorial bulge, the Earth's axis wobbles slowly. It stays tilted, but the direction of the tilt makes a slow circle once every 25,780 years. This is called "the precession of the equinoxes."

About 12,890 years from now, the axis will be tilting in the opposite direction, so that if that is the only change, the summer solstice will come on December 21, and the winter solstice on June 21. The summer solstice would then be at perihelion and the northern summer would be hotter than it is now. The winter solstice would be at aphelion and the northern winter would be colder than it is now. In other words, the situation would be the reverse of what it is at present. The northern hemisphere would have cold winters and hot summers while the southern hemisphere would have mild winters and cool summers.

But there are other factors. The perihelion point is slowly moving about the sun. Every time the Earth travels about the sun, it reaches the perihelion point at a slightly different place and time. The perihelion (and aphelion, too) makes a complete circle about the sun in about 21,310 years. Every 58 years, the day of perihelion shifts by one day in our calendar.

But that's still not all. One of the effects of the various gravitational pulls on Earth is to cause the axial tilt to wobble in actual amount. Right now, the axial tilt is 23.44229° but in 1900, it was 23.45229° and in 2000 it will be 23.43928°. As you see, the axial tilt is decreasing, but it will only decrease so far and then it will increase again, then decrease, and so on. It never gets less than about 22° and it never gets more than about 24.5°. The length of the cycle is 41,000 years.

A smaller tilt of the axis means that both the northern and southern ends of the Earth get less sun in the summer and more in the winter. The result is milder winters and cooler summers for *both* hemispheres. Contrariwise, the larger the axial tilt, the more extreme the seasons for *both* hemispheres.

Finally, Earth's orbit gets more and less eccentric. The eccentricity, which is at this time 0.01675, is decreasing and will eventually reach a minimum value of 0.0033, or only 1/5 its present amount. At that time the Earth will be only 990,000 kilometers (610,000 miles) closer to the sun at perihelion than at aphelion. Afterward, the eccentricity will start increasing again to a maximum of 0.0211, or 1.26 times the present value. Then the Earth will be 6,310,000 kilometers (3,920,000 miles) closer to the sun at perihelion than at aphelion. The less the eccentricity and the more nearly circular the orbit, the smaller the difference in the amount of heat the Earth gets from the sun at different times of the year. This encourages the mild-winter/cool-summer situation.

If all these variations in the Earth's orbit and its axial tilt are taken into consideration, it would seem that, on the whole, the tendency to mild seasons and to extreme seasons alternates in a roughly 100,000-year cycle.

In other words, each of Milankovich's "Great Seasons" lasts about 25,000 years. We seem now to have passed the "Great Spring" of the retreating glaciers, and will continue to pass through the Great Summer and Great Fall, into the Great Winter of an ice age about 50,000 years from now.

Is all this theorization correct, though? The variations in the orbit and in the axial tilt are small and the difference between the cold-winter/hot-summer and mild-winter/cool-summer is not really great. Is the difference enough?

The problem was tackled by three scientists, J. D. Hays, John Imbrie, and N. J. Shackleton, and their results were published in December 1976. They worked on long cores of sediment dredged up from two different places in the Indian Ocean. The places were far from land areas so there would be no material washed from the coastline to obscure the record. The places were also relatively shallow so that there would be no material washed down from surrounding, less deep areas.

The sediment, it could be supposed, would be undisturbed material laid down on the spot for century after century, and the length of the core brought up stretched backward, it seemed, over a period of 450,000 years. The hope was that there would be changes as one went along the cores, changes that would be as distinctive as the changes in tree rings that enabled one to differentiate dry summers from wet summers.

One change was in connection with tiny Radiolaria that lived in the ocean through all the half-million years being investigated. These are one-celled protozoa with tiny, elaborate skeletons that, after death, drift down to the sea bottom as a kind of ooze. There are numerous species of Radiolaria, some of which flourish under warmer conditions than others. They are easily distinguished from each other by the nature of their skeletons, and one can therefore creep along the sediment cores, millimeter by millimeter, studying the nature of the radiolarian skeletons and estimating from them whether, at each given time, the ocean water was warm or cool. One could, in this way, set up an actual curve of ocean temperature with time.

One can also with time follow the change in ocean temperature, by noting the ratio of two varieties of oxygen atoms: oxygen-16 and oxygen-18. Water containing oxygen-16 in its molecules evaporates more easily than water containing oxygen-18.

That means that rain or snow falling on land is made up of molecules richer in oxygen-16 and poorer in oxygen-18 than is ocean water. If a great deal of snow falls on land and is tied up in glaciers, the ocean water that is left suffers a considerable deficit of oxygen-16 while the oxygen-18 piles up.

Both systems for judging the temperature of the water (and the prevalance of ice on land) gave identical results, even though they were widely different in nature. What's more, the cycle produced by these systems was very much like the cycle calculated from the changes in Earth's orbit and in its axial tilt.

It would therefore seem, at the moment, and pending further evidence, that the Milankovich notion of the Great Seasons looks good.

The Arctic Ocean

If the ice ages follow the Great Seasons, then we should be able to predict precisely when the next ice age will start. It should be about 50,000 years from now.

Of course, we needn't suppose that the cause of ice ages is unitary in nature. There may be more than one contributory cause. For instance, the orbital and axial changes may set up the basic period, but other effects ought to have an influence and a less regular one. Changes in the sun's radiation, or in the dustiness of space between Earth and sun, or in the carbon dioxide content of the atmosphere, may, singly or together, affect the cycle, reinforcing it on some occasions and counteracting it on others.

If two or more effects coincide, an ice age might be worse than usual. If the orbital and axial changes are counteracted by unusually clear space, or an unusually high carbon dioxide content, or an unusually spotty sun, an ice age might be rather mild or might be skipped altogether.

In the present case we might fear the worst since in 50,000 years we will not only reach the Great Winter, but we may also (as I said earlier in the chapter) be entering a cosmic cloud which will cut down the solar radiation reaching us.

Yet we might still be completely off-base. After all, the orbital-axial swings should have been continuing with absolute regularity for as long as the solar system has existed in its present structure. There should have been ice ages every hundred thousand years or so all during the history of life.

Instead, ice ages have been matters of only the last million years or so. Before that, for a period of about 250 million years, there seem to have been no ice ages at all. It may even be that there are successive periods of ice ages over a couple of million years separated from each other by intervals of a quarter of a billion years.

But why the intervals? Why were there no ice ages during those long intervals, when the orbital-axial swings continued in those intervals exactly as they do now. The reason may lie in the land-sea configuration on Earth's surface.

If a polar region consisted of a vast expanse of sea, there would be some millions of square kilometers of sea ice, not very thick, swirling about the pole. The sea ice would be thicker and more extensive in winter, thinner and less extensive in summer.

During the ice age end of the orbital-axial swing, the sea ice would be, on the whole, thicker and more extensive winter and summer, but not very much more so. After all, there are ocean currents that continually bring warmer water up from the temperate and tropical regions and this tends to ameliorate polar weather, even during an ice age.

Again, if a polar region consisted of a continent with the pole more or less in its center and with unbroken sea surrounding it, we would expect the continent to be covered with a thick ice cap that would not melt during the very cool summer and that would accumulate from year to year.

The ice wouldn't accumulate forever, of course, since ice under considerable weight flows—as Agassiz proved a century and a half ago. The ice gradually flows into the surrounding ocean, breaking off as huge icebergs. The icebergs, together with sea ice, would float around the polar continent and would gradually melt as they drift toward more temperate latitudes. In an ice age, the icebergs would multiply and in interglacial periods they would diminish, but the change would not be great. The surrounding ocean, thanks to ocean currents, would maintain its temperature very close to normal, ice age or not.

Such a case really exists on Earth, for Antarctica is covered by a thick ice cap and the ocean surrounding it is choked with ice. Antarctica, however, has had that ice cap for some 20 million years, and has scarcely been affected by the coming and going of the ice ages.

Suppose, however, that you have a polar ocean, but *not* a vast one.

Instead, suppose you have a small, nearly landlocked one, like the Arctic Ocean. The Arctic Ocean, no larger than the continent of Antarctica, is almost entirely surrounded by the huge continental masses of Eurasia and North America. The only considerable connection between the Arctic Ocean and the rest of the waters of the world is a strait 1,600 kilometers (1,000 miles) wide between Greenland and Scandinavia and even this is partially blocked by the island of Iceland.

It is the northern land that makes all the difference. During the triggering of an ice age, the additional snow that would fall during a mild winter would fall on land and not on the ocean. On the ocean, the snow would simply melt, because water has a high heat capacity, and because even if the accumulating snow were capable of lowering the temperature of the ocean to the point of freezing, water currents from warmer climes would prevent that.

On land, however, the snowflakes have a better chance. Land has a lower heat capacity than water so it cools down much more rapidly under a given load of snow. Furthermore, there are no currents of any kind to ameliorate the actions so that the ground freezes hard. Then, if there is not enough heat in the summer to melt all the snow, the snow turns to ice and the glaciers begin their march.

The presence of the large land areas circling the North Pole provides a huge receiving area for the snow and ice, and the Arctic Ocean (particularly before the advance of the Ice Age covered it with sea ice) provides the water source. The ocean-continent arrangement is just right in the northern hemisphere to maximize the Ice Age effect.

But the ocean-continent arrangement in the northern hemisphere is not a permanent thing. It constantly changes as a result of plate tectonics.

It follows, then, that as long as the Earth's surface is so arranged that the polar regions are either open ocean, or are isolated continents surrounded by open ocean, there are no spectacular ice ages. It is only when the moving plates happen to bring about an arrangement as exists in the north polar regions today that the orbital-axial cycle brings the kind of ice ages we are now familiar with. That happens, apparently, only once in 250 million years or so.

But we are there now and certainly the arrangement of the continents won't change dramatically for another million years or so, so that we are due not only for another ice age but for a whole series of them.

The Effect of Glaciation

Suppose an ice age does come. How bad a disaster might it be? After all, we've had a million years of glaciers coming and going and here we all are That's true and, if we stop to think of it, the glaciers creep slowly. They take thousands of years to advance and even at the stage of maximum glaciation, it is surprising how little change important parts of the world undergo.

Right now, there are some 25 million cubic kilometers (6 million cubic miles) of ice resting on various land surfaces of the world, chiefly Antarctica and Greenland. At the height of glaciation there was a monster ice sheet covering the northern half of North America and smaller ice sheets in Scandinavia and northern Siberia. At that time, a total of perhaps 75 million cubic kilometers (18 million cubic miles) of ice rested on land. This means that at the height of glaciation, 50 million cubic kilometers (12 million cubic miles) of water that is now in the ocean was then on land.

The water subtracted from the ocean to feed the glaciers was, however, even at the height of glaciation, only 4 percent of the total. This means that even at the height of the glaciation, 96 percent of the ocean was right where it is now.

From the standpoint of sheer room, therefore, sea life would feel no particular constriction of the environment. To be sure, the ocean water would, on the average, be somewhat colder than it is now—but what of that? Cold water dissolves more oxygen than warm water does, and sea life depends on oxygen as much as we do. That is why the polar waters are far richer in life than the tropic waters and why the polar waters can support giant mammals that live on sea animals —such as great whales, polar bears, elephant seals, and so on.

If, during an ice age, the ocean water is colder than it is now, that would actually encourage life. It might be *now* that sea life feels the pinch, not then.

The situation would be different on land, and there it might seem that matters were much more disastrous. At the present moment, 10 percent of the Earth's land surface is covered with ice. At the height of a glaciation, that amount was tripled; 30 percent of the Earth's present land surface was covered by ice. This meant that the area open to land life was reduced from about 117 million square kilometers (45 million square miles) of land that was ice-free at least in the summer, to a mere 90 million square kilometers (35 million square miles). Yet, that is not a fair description of what actually happens.

At the height of a glaciation, the loss of 4 percent of the ocean's liquid water meant a drop in sea level of as much as 150 meters (490 feet). That doesn't change the ocean itself much, but around each continent are tracts of land that are under only shallow depths of the ocean's rim. These sections with less than 180 meters (590 feet) of water above them are called the "continental shelves." As the sea level drops, most of the continental shelves are little-by-little exposed and are open to the invasion of land life.

In other words, as the glaciers advance and swallow up land, the sea level drops and exposes new land. The two effects may largely balance. Since the glaciers advance with extreme slowness, vegetation drifts slowly southward and onto the exposed continental shelves ahead of the glaciers, animal life naturally following the vegetation.

As the glaciers advance, the stormbelts retreat southward, too, bringing rain to the warmer parts of the Earth that were unused to it before (and since). In short, what are now deserts were not deserts during the Ice Age. Before the last retreat of the glaciers, what is now the Sahara Desert consisted of fertile grasslands.

We might argue that with the exposure of continental shelves and the shrinkage of deserts, the total land area exposed to a rich saturation with life forms was greater at the height of an ice age than it is right now, paradoxical though it may seem. In particular, during the last Ice Age, human beings—not our hominid ancestors, but *Homo sapiens* itself—moved south as the glaciers advanced, north as they retreated, and flourished.

How, then, would an ice age be different in the future? For instance, suppose the glaciers were starting a new advance now. How disastrous would this be?

To be sure, humanity is less mobile now than it used to be. At the time of the last Ice Age, there may have been 20 million human beings on Earth altogether; now, there are 4 billion, two hundred times as many. It is more difficult for 4 billion people to move than for 20 million.

Then, too, consider the change in life-style. At the time of the last Ice Age, human beings were in no way bound to the soil. They were food gatherers and food hunters. They followed the vegetation and the animals, and all places were alike to them as long as they could find fruits, nuts, berries, and game.

Since the last Ice Age, human beings have learned to be farmers and miners. Farms and mines cannot be moved. Nor can the vast structures that human beings have built, the cities, tunnels, bridges, power lines, and so on, and so on, and so on. None of that can be moved; it can only be abandoned and new items built elsewhere.

Never forget, though, how slowly the glaciers advance and retreat, and how slowly the sea level sinks and rises as a result. There will be plenty of time to make the shift, nondisastrously. We can imagine humanity slowly moving south and onto the continental shelves—then into the interior and north again—over and over in a slow alternation while the present continental configuration about the North Pole lasts. It would be a kind of 50,000-year expiration, followed by a 50,000-year inspiration, and so on.

It wouldn't be a steady movement, for the glaciers advance with intervals of partial retreat, and they retreat with intervals of partial advance; but human beings, with difficulty, will imitate those advances and retreats in all their complexities—provided all the advances and retreats are slow enough.

To be sure, the differences in environment do not necessarily involve only the advance in glaciers. The retreat of the glaciers since the last Ice Age is not absolute. There remains the ice cap in Greenland, an unmelted relic of the Ice Age. What if, with a Great Summer ahead of us, the climate continued to mellow and the north polar ice melted, including even the Greenland ice cap.

The Greenland ice cap contains 2.6 million cubic kilometers (620,000 cubic miles) of ice. If that and the lesser ice sheets on some of the other polar islands were to melt and pour into the ocean, sea level would rise some 5.5 meters (17.5 feet). That would be an embarrassment to some of our coastal areas, and particularly low-lying cities, such as New Orleans, would be flooded out. Again, if the melting took place slowly enough, and the sea level rose slowly enough, we could imagine the coastal cities slowly abandoning the shoreline and retreating to higher grounds, nondisastrously.

Suppose that for some reason the Antarctica ice sheet also melted. It is not likely to in the natural course of things, for it has survived all the interglacial periods of the past—but suppose! Since 90 percent of Earth's ice supply rests on Antarctica, if that melted the sea level would rise by ten times the amount that the melting of Greenland would make possible. The sea level would rise by some 55 meters (175 feet) and the water would reach the eighteenth story of the New York skyscrapers. The low-lying borders of the present continents would be under water. The state of Florida, and many of the other

Gulf states would be gone. So would the British Isles, the Netherlands, north Germany, and so on.

However, Earth's climate would become much more equable, and there would be neither polar lands nor desert lands. Again, the room available for humanity might remain as large as before and if the change were slow enough, even the melting of Antarctica would not be terribly disastrous.

If the coming of the next ice age or of the melting of Antarctica is postponed for some tens of thousands of years, however, none of it may happen. Humanity's advancing technology may well be able to modify the ice age trigger and keep the Earth's average temperature in place, if that is wished.

For instance, mirrors may be placed in near space, which may be adjustable and which may serve to reflect sunlight that would ordinarily miss Earth, onto Earth's night surface; or it may reflect sunlight that would ordinarily hit Earth's day surface and keep it from reaching Earth at all. In this way, Earth might be warmed slightly if the glaciers threaten, or cooled if ice-melting threatens.*

Again, we may develop methods for altering the carbon dioxide content of the Earth's atmosphere in a controlled manner, thus allowing more heat to escape from Earth if ice-melting threatens, or conserve the heat if the glaciers threaten.

Finally, as more and more of Earth's population swarms into space settlements, the comings and goings of the glaciers will become less important to humanity as a whole.

In short, ice ages as they have occurred in the past would not be catastrophic in the future, and may not even be disastrous. In fact, they may not even occur, thanks to human technology.

But what if the glaciers approach unexpectedly and at unprecedented speeds, or what if Earth's ice supply melts unexpectedly and at unprecedented speeds—and what if this happens before we are technologically ready for it. *Then* we might experience a huge disaster and even a near-catastrophe and there are conditions under which this might happen—something I will take up later.

---

* A similar device may serve to keep Earth habitable for some tens of thousands of years, perhaps, after the gradually warming sun would ordinarily have made it uninhabitable—if people bother to take the trouble.

Chapter 11

The Removal
of Magnetism

Cosmic Rays

Although the various disasters in which Earth has been involved, from ice ages to earthquakes, have never been enough to wipe out life upon the planetary surface, as Cuvier and the catastrophists imagined a century and a half ago, there have been near-catastrophes—occasions when life has suffered some devastating damage. At the close of the Permian period, 225 million years ago, over a comparatively short period of time about 75 percent of the families of amphibians and 80 percent of the families of reptiles that had been alive in the Permian came to an end. It was an example of what some people have come to call "a great dying."

Six times since then there seem to have been such great dyings. The time most commonly referred to in that manner came at the end of the Cretaceous period about 70 million years ago. At that time, the

213

dinosaurs, after having flourished for 150 million years, died out completely. So did the large sea reptiles such as the ichthyosaurs and plesiosaurs, and the flying pterosaurs. Among the invertebrates, the ammonites, which had been a large and flourishing group, died out. In fact, as many as 75 percent of all the animal species then living may have become extinct over a comparatively short period.

It seems likely that such great dyings must have been the result of some marked and comparatively sudden change in the environment; but it must have been a change that left large numbers of species still alive and, as nearly as we can tell, scarcely affected.

A particularly logical explanation involves the shallow seas which now and then invade the continents, and now and then drain away. The invasion may take place when the ice load on polar land areas is particularly low; and the draining may take place during periods of mountain building when the average altitude of continents rises. In any case, shallow inland seas offer favorable environments for large numbers of species of sea animals and these, in turn, offer a stable and copious food supply for other animals that live on the shores. When the inland seas drain away, both the sea animals themselves and the land animals that live on them naturally die out.

In five of the seven cases of great dyings in the last quarter of a billion years, there seem to have been periods of sea drainage. The explanation also fits the fact that marine animals seem more subject to great dyings than land animals, and that the plant world seems scarcely affected at all.

Although sea drainage may be the most logical and reasonable solution to the problem (and one that holds no terrors for human beings who do not live in inland seas and who live in a world in which there are no important inland seas) many other suggestions have also been offered to explain the great dyings. One of those explanations, while not perhaps very likely, is unusually dramatic. What's more, it introduces a type of catastrophe we have not yet considered, and one which may be threatening to humanity. It involves radiation from space that does not come from the sun.

In the early years of the twentieth century, radiation was detected that was even more penetrating and energetic than the newly discovered radiations arising from radioactive atoms. In 1911, in order to make sure that this penetrating radiation was coming from the ground, the Austrian physicist Victor Francis Hess (1883–1964) sent radiation-detecting devices as much as 9 kilometers (5.6 miles) into the air in balloons. He expected to find the level of radiation decreasing because so much of it would be absorbed by the air between the ground and the balloon.

It turned out, on the contrary, that the intensity of penetrating radiation increased with height so that it was clear that it was coming from the outside universe, or cosmos. Hence the name "cosmic rays" was given to the radiation in 1925 by the American physicist Robert Andrews Millikan (1868–1953). In 1930, the American physicist Arthur Holly Compton (1892–1962) was able to show that the cosmic rays were very energetic, positively charged particles. We now understand how cosmic rays originate.

The sun, and presumably every star, undergoes processes that are energetic enough to spray particles into space. These particles are, for the most part, atomic nuclei. Since the sun is mostly hydrogen, hydrogen nuclei, which are simple protons, are the most common particles involved. Other, more complex nuclei, occur in minor quantities.

These energetic protons and other nuclei stream out from the sun in every direction and are the solar wind to which I have referred earlier.

When the sun undergoes particularly energetic events, the particles are hurled outward with greater energy. When the sun's surface erupts in large "flares," very energetic particles are included in the solar wind and these may reach the lower limits of the energies associated with cosmic rays. (They are referred to as "soft cosmic rays.")

Other stars send out stellar winds and those stars which are more massive and hotter than the sun send out more energetic winds richer in particles at cosmic ray energy levels. Supernovas in particular send out vast floods of energetic cosmic rays.

The particles of the cosmic rays, being electrically charged, follow a curved path when passing through a magnetic field. Every star has a magnetic field and the Galaxy as a whole has one. Each cosmic ray particle therefore follows a complex curved path and, in the process, is accelerated by the magnetic fields it passes through and gains still more energy.

In the end, all of interstellar space within our galaxy is rich in cosmic ray particles flying in every direction according to how the twists and turns of the magnetic field through which they have passed have directed them. A certain tiny percentage of these are bound by sheer chance to strike the Earth, and they do so from every possible direction.

Here, then, we have a new kind of invasion from outer space that we have not yet considered. Earlier, I pointed out how unlikely it was that the solar system encounter another star or that it be penetrated by even small bits of matter originating from other planetary systems. Later, I mentioned the dust particles and atoms of the interstellar clouds.

Now we must consider the invasions from the space beyond the solar system of the smallest of all material objects, subatomic particles. There are so many of these, and they are distributed so thickly through space, and travel at speeds so close to that of light, that Earth is under constant bombardment by them.

Cosmic rays make no visible mark upon the Earth, however, and we are not aware of their coming. Only scientists with their special detecting devices can be aware of the cosmic rays and that only in the last two generations.

Furthermore, they have been falling on the Earth throughout planetary history and life on this planet does not seem to have been any the worse for it. Nor have human beings seemed to suffer from it in the course of their history. It would seem, therefore, that we can eliminate it as a source of catastrophe—and yet we can't.

To see why that is, let us dive into the cell.

DNA and Mutations

Every living cell is a tiny chemical factory. The properties of a particular cell, its shape, its construction, its abilities, depend on the exact nature of the chemical changes that go on within it, the rate at which each proceeds, and the manner in which they are all interrelated. Such chemical reactions would usually proceed very slowly and even imperceptibly, if the substances making up the cells and participating in the reactions were simply mixed together. In order for the reactions to proceed at a rapid and smoothly regulated rate (as they are observed to do and as is necessary if the cell is to live) those reactions must take place with the aid of certain complex molecules called "enzymes."

Enzymes are members of a class of substances called "proteins." Proteins are made up of giant molecules each constructed of chains of smaller building blocks called " amino acids." Those amino acids come in some twenty varieties and are capable of being put together in any conceivable order.

Suppose we begin with one each of those twenty amino acids and put them together in all possible orders. It turns out that the total number of different orders in which they can be put together is equal to about 50,000,000,000,000,000,000 (fifty billion billion), with each different order representing a distinctly different molecule. Actual enzyme molecules are made up of a hundred or more amino acids and the number of possible ways of combining those amino acids is unimaginably enormous. Yet a particular cell will only contain a certain limited number of enzymes, with each molecule of a particular enzyme construction of an amino-acid chain made up of amino acids in one specific order.

A particular enzyme is so constructed that particular molecules will attach themselves to the enzyme surface in such a way that interaction between them—involving a transfer of atoms—can very readily take place. After interaction, the changed molecules will no longer hold to the surface. They move off and other molecules attach themselves and undergo the reaction. As the result of the presence of even a few molecules of a particular enzyme, large numbers of molecules react with each other that would not have reacted at all if the enzyme had not been there.*

What it boils down to, then, is that the shape, construction, and abilities of a particular cell depends on the nature of the different enzymes in that cell, the relative numbers of the different enzymes, and the manner in which they do their work. The properties of a multicellular organism depend on the properties of the cells that make it up and on the way those individual cells interrelate. In the long run, then (and in no very simple way, of course), all organisms, including human beings, are the product of their enzymes.

But this seems a very chancy dependence. If an enzyme is not constructed of a precise order of amino acids, it may not be able to do its work. Change one amino acid into another and the enzyme surface may not serve as a proper "catalyst" for the reaction it controls.

What is it then that forms the enzymes so precisely? What is it that sees to it that one particular amino acid order is set up for a particular enzyme, and no other order? Is there some key substance in the cell which carries a "blueprint," so to speak, of all the enzymes in the cell, thus guiding their manufacture?

---

* The situation is rather like that of tossing a needle and thread separately into the air and hoping they will thread themselves by chance; or of holding a needle in one hand, a thread in the other and deliberately threading the needle. The former would be like a cellular reaction without an enzyme, and the latter the same reaction with an enzyme.

If there is such a key substance, it must be in the "chromosomes." These are little objects within the central nucleus of a cell and they *behave* as though they carry the blueprint. The chromosomes come in varying numbers in different species of organisms. In the human being, for instance, each cell contains twenty-three pairs of chromosomes.

Every time a cell divides, each chromosome divides first into two chromosomes, each the replica of the other. In the process of cell division, one of the replicas of each chromosome goes into one cell while the other replica goes into the other cell. In this way, each daughter cell ends up with twenty-three pairs of chromosomes, the two sets of pairs being identical. This is what you would expect if the chromosomes carried the blueprint for enzyme structure.

All but the most primitive organisms develop sex cells whose task it is to form new organisms in a more complicated way than by simple cell division. Thus, male human beings (and the males of most other complex animals) produce sperm cells, while females produce egg cells. When a sperm cell joins or "fertilizes" an egg cell, the resultant combination can undergo repeated divisions until a free-living new organism is formed.

Both egg cells and sperm cells have only half the usual number of chromosomes. Each egg cell and each sperm cell gets only one of each of the twenty-three pairs. When they combine, the fertilized egg has twenty-three pairs of chromosomes again, but one of each pair comes from the mother and one from the father. Thus, the offspring inherits characteristics equally from its two parents and the chromosomes behave exactly as you would expect if they carried the blueprint for enzyme-manufacture.

But what is the chemical nature of this supposed blueprint?

Ever since the discovery of chromosomes in 1879 by the German anatomist Walther Flemming (1843–1905), there was the general assumption that the blueprint, if it existed, would have to be a very complex molecule, and that meant it would have to be a protein. The proteins were the most complicated substances known to exist in tissue, and the enzymes, it was discovered in 1926 by the American biochemist James Batchellor Sumner (1887–1925), were themselves proteins. Surely it should be a protein that serves as a blueprint for the construction of other proteins.

In 1944, however, the Canadian physician Oswald Theodore Avery (1877–1955) was able to show that the blueprint molecule was not protein at all, but was another type of molecule called "deoxyribonucleic acid," or DNA for short.

This was a complete surprise, for DNA was thought to be a simple molecule, one that was not at all suitable to serve as blueprint for the complex enzymes. Closer examination, however, showed that DNA was a complex molecule; more complex, indeed, than proteins were.

Like the protein molecule, the DNA molecule was made up of long chains of a simpler building block. In the case of DNA the building block was called a "nucleotide" and an individual DNA molecule could be built up of chains of many thousands of nucleotides. The nucleotides consisted of four different varieties (not twenty, as in the case of proteins) and these four varieties could be hooked together in any order whatever.

Suppose you took the nucleotides three at a time. There would then be 64 different "trinucleotides." If you number the nucleotides 1, 2, 3, and 4, you could have the trinucleotides: 1-1-1, 1-2-3, 3-4-2, 4-1-4, and so on, 64 different combinations. One or more of these trinucleotides could be equivalent to a particular amino acid; some might indicate "punctuation," such as starting an amino-acid chain, or ending it. The translation of the trinucleotides of the DNA molecule into the amino acids of the enzyme chain is called the "genetic code."

But that simply seems to remove the difficulty one stage further back. What enables the cell to construct a particular DNA molecule that will lead to the construction of a particular enzyme molecule— out of all the uncountable number of different DNA molecules that could exist?

In 1953, the American biochemist James Dewey Watson (1928–) and the English biochemist Francis H. C. Crick (1916–) were able to work out the structure of DNA. It existed in two strands coiled into a double helix. (That is, each strand had the shape of a spiral staircase, and the two strands were interlaced.) Each strand was, in a way, the opposite of the other, so that they fit together neatly. In the process of cell division, each DNA molecule uncoiled into two separate strands. Each strand then brought about the construction of a second strand on itself, one that fit it snugly. Each strand served as a blueprint for the new partner and the result was that where one double-helix had originally existed, two double-helixes were formed, each the replica of the other. The process was called "replication." Thus, once a particular DNA molecule existed, it propagated itself, retaining its exact configuration, from cell to daughter-cell and from parent to offspring.

It follows that every cell and, indeed, every organism right up to the human being, has its shape, its structure, and its chemistry (and to a certain extent, even its behavior) determined by the exact nature of its DNA content. The fertilized egg of one species of organism doesn't look very different from that of another, but the DNA molecules in each are completely distinct. For that reason a human fertilized egg will develop into a human being and a giraffe fertilized egg will develop into a giraffe, and no confusion between the two is possible.

But, as it happens, the transmission of DNA molecules from cell to daughter-cell and from parent to offspring is not as perfect as all that. It is the experience of herdsmen and farmers that every once in a while, young animals or plants are produced that do not have quite the characteristics of the parent organisms. Generally, these differences are not great and sometimes not even particularly noticeable. Occasionally, an aberration is so extreme as to produce what is called a "sport" or "monster." The scientific term for all such offspring with changed characteristics, extreme or unnoticeable, is "mutation," from a Latin word meaning "change."

Generally, pronounced mutations were viewed with uneasiness and destroyed. In 1791, however, a Massachusetts farmer named Seth Wright took a more practical view of a sport that turned up in his flock of sheep. A lamb was born with abnormally short legs, and it occurred to the shrewd Yankee that short-legged sheep could not escape over the low stone walls around his farm. He therefore deliberately bred a line of short-legged sheep from this not entirely unfortunate accident and this helped bring mutations to the attention of people generally. It was not, however, until the work of the Dutch botanist Hugo Marie de Vries (1848–1935) in 1900 that mutations came to be studied scientifically.

Actually, where mutations were not very pronounced and therefore not frightening or repellent, herdsmen and farmers routinely took advantage of them. By selecting from each generation those animals that seemed most suitable to human exploitation—cows that gave much milk, hens that laid many eggs, sheep that grew much wool, and so on—strains were developed that differed greatly from each other and from the original wild organism that had first been tamed.

This is the result of selecting small and not, in themselves, very important mutations which, however, like Wright's short-legged sheep, pass on the mutation to their offspring. By choosing mutation after mutation, all in the same direction, the strains are "improved" from the human standpoint. We have but to think of the numerous strains of dogs or of pigeons to realize how neatly we can shape and form a species by carefully directing matings and by saving some offspring and discarding others.

The same can be done even more easily with plants. The American horticulturalist Luther Burbank (1849–1926) made a successful career of breeding hundreds of new varieties of plants which were improvements over the old in one respect or another, not only by mutations, but by judicious crossing and grafting.

What human beings do purposely, the blind forces of natural selection do very slowly over the course of the ages. In every generation, the offspring of a particular species vary from individual to individual in part because of slight mutations that take place. Those whose mutations allow them to play the game of life more efficiently, have a greater chance of surviving and passing on those mutations to more numerous offspring. Little by little over millions of years, new species are molded out of old ones, one species replaces another, and so on.

This was the essential core of the theory of evolution by natural selection advanced in 1858 by the English naturalists Charles Robert Darwin and Alfred Russel Wallace.

On a molecular level, mutations are the result of imperfect replications of DNA. These can take place from cell to cell in the process of cell division. In that case, within an organism a cell can be produced which is not like the other cells of the tissue. This is a "somatic mutation."

Generally, a mutation is for the worse. After all, if we consider an intricate molecule of DNA replicating itself and getting a wrong building block into position, it isn't likely to do a better job because of the mistake. The result is, then, that a mutated cell within the skin or liver or bone is going to work so poorly that it will be virtually out of action and very likely unable to multiply. Other normal cells about it will continue to multiply when necessary and will crowd it out of existence. Thus the tissue as a whole stays normal despite occasional mutations.

The major exception to this is when the mutation just happens to affect the process of growth. Normal cells in a tissue grow and divide only as needed to replace lost or damaged cells, but a mutated cell may lack the mechanism designed to stop the growth at the right time. It may simply grow and multiply helplessly, regardless of the needs of the whole. Such an anarchic growth is cancer, and it is the most serious result of a somatic mutation.

Occasionally, a DNA molecule will mutate in such a way as to do a better job under a certain condition. This won't happen often, but cells containing it will flourish and survive, so that natural selection works not only on organisms as a whole, but on the DNA blueprints as well, and that is how the first DNA molecules must have been formed—from the simple building blocks by random factors until one was built up capable of replication, and evolution did the rest.

Every once in a while, sperm cells or egg cells are formed with imperfectly replicated DNA. These produce mutated offspring. Again, most mutations are for the worse and the mutated offspring are either incapable of developing, or die young, or even if they live on and have offspring they are gradually overrun by more efficient individuals. Very occasionally a mutation happens to be for the better under a given set of conditions, and that mutation is likely to establish itself and flourish.

Although mutations for the better occur much less frequently than mutations for the worse, it is the former that tend to survive and to crowd out the latter. For that reason, anyone watching the course of evolution, can imagine there is purpose behind it—as though organisms are deliberately trying to better themselves.

It is hard to believe that random processes, hit-and-miss, can produce the results we see about us today—but given enough time and given a system of natural selection which allows millions of individuals to perish so that a few improvements may establish themselves, random processes will do the job.

The Genetic Load

But why do DNA molecules replicate themselves imperfectly now and then? Replication is a random process. As nucleotide building blocks line up against a DNA strand, only the one particular nucleotide that fits should ideally form against each particular nucleotide of the already existing strand. Only it would stick, so to speak. Members of the other three nucleotides would not.

Yet, by the blind movement of molecules, a wrong nucleotide might strike a particular nucleotide on the strand and, before it could bounce off, be locked in on either side by other nucleotides that had fitted themselves with too great an efficiency. Now you would have a new DNA strand that is not exactly what is called for, but which differs by one nucleotide and will, in consequence, produce an enzyme that differs in one amino acid. Nevertheless, the imperfect strand forms a new model in future replications and serves to replicate itself and not the grand original.

Under natural circumstances, the chance of an imperfect replication of a particular DNA strand on a particular occasion is only 1 in 50,000 to 100,000, but there are so many genes in living organisms and there are so many replications that the chance of a mutation now and then amounts to a certainty. There are many of them.

It is possible that, among human beings, as many as 2 in 5 fertilized egg cells contain at least one mutated gene. This means that some 40 percent of us are mutants in one way or another with respect to our parents. Since mutated genes tend to be passed through the generations for a period of time before dying out, some estimates are that individual human beings carry an average of eight mutated genes—in almost every case, genes that are mutated for the worse. (That this doesn't affect us more than it does is due to the fact that genes occur in pairs, and where one gene is abnormal, the other will carry us through.)

Nor are the chances of mutation necessarily due to blind chance alone. There are factors that tend to increase the chance of imperfect replication. There are various chemicals, for instance, that interfere with the smooth working of DNA and hamper its ability to work with the proper nucleotides only. The chances of mutations then obviously increase. Since the DNA molecule is a very intricate and delicate structure, many chemicals can interfere with it. Such chemicals are called "mutagens."

Then, too, there are subatomic particles that will do the trick. The DNA molecules are hidden in the chromosomes which are themselves buried in the nuclei in the center of the cells and chemicals have some difficulty in getting at them. Subatomic particles, however, smash through the cells and, if they hit the DNA molecules, can knock atoms out of their structure and change them physically.

DNA molecules can be so mishandled in this fashion as to be unable to replicate at all, and the cell can be killed. If a large number of critical cells are killed, the individual will die of "radiation sickness."

Less drastically, the cell may not be killed but a mutation may be produced. (The mutation may be cancer-producing and it is known that energetic radiation is "carcinogenic"—cancer-producing—as well as mutagenic. In fact, one implies the other.) Of course, if the egg cells or sperm cells are affected, offspring with mutations are produced, sometimes so drastic that serious birth-defects are observed. (This can be brought about by chemical mutagens as well.)

The mutagenic effect of radiations was first demonstrated in 1926 by the American biologist Hermann Joseph Muller (1890–1967), when he studied mutations in fruit flies more easily by deliberately bringing about increased numbers of them when he exposed the insects to X-rays.

X-rays and radioactive radiations were not known to man, and were therefore not produced, before the twentieth century, but that doesn't mean that there were no mutagenic forms of radiation prior to that. There has always been sunlight as long as life has existed on Earth, and sunlight is weakly mutagenic because of the ultraviolet light it contains (and, because of that, overexposure to sunlight results in an increased chance of developing skin cancer.)

Then, too, there are the cosmic rays to which life has been exposed as long as it has existed. Indeed, one might argue (though others disagree) that cosmic rays, through the mutations they induce, have been the chief driving force behind evolution over the last few billion years. Those eight mutated genes per individual—almost all deleterious ones—are the price we pay, so to speak, for the few beneficial ones here and there on which the future depends.

Of course, if a little is good, that does not mean a lot is better. The mostly deleterious mutations that are produced by whatever cause represent a debilitating effect on a given species, since they result in a number of substandard individuals. This is the "genetic load" for that species (a term first introduced by H. J. Muller). There are still, however, a substantial percentage of individuals without seriously bad mutations, together with a few individuals possessing a beneficial mutation. These manage consistently to outsurvive and outbreed the substandards so that, as a whole, a species survives and progresses despite its genetic load.

But what if the genetic load increases because the rate of mutation increases for some reason? That would mean more substandard individuals and fewer normal or superstandard ones. Under such conditions, there may simply not be enough normal and superstandard individuals to keep the species growing in the face of all the substandards. In short, increasing the genetic load will not speed up evolution as one might think—but will weaken the species and lead to extinction. A small genetic load has its uses; a large one is deadly.

But what can cause an increase in the mutation rate? Random factors remain random and most of the mutagenic factors in past history—sunlight, chemicals, natural radioactivity—have been more or less constant in their influence. But what about cosmic rays? What if, for some reason, the cosmic ray intensity reaching Earth increases? Might that weaken many species and lead to a great dying through genetic loads that become too heavy to survive?

Even if we were to agree that the actual great dyings in Earth's history were to be attributed to the draining of inland seas, might it be that a sudden increase in cosmic ray intensity could *also* result in a great dying? Possibly, but what would cause a sudden increase in cosmic ray intensity?

One possible cause is an increase in the incidence of supernovas which are, after all, the prime sources of cosmic rays. That probably is not very likely. In the hundreds of billions of stars in our galaxy, the total number of supernovas from year to year and century to century is likely to stay about the same. Might it be that the distribution of supernovas changes; that at some times a disproportionate number of them are at the other end of the Galaxy and at other times a disproportionate number are at our end?

Actually, this would not affect the cosmic ray intensity as much as we might think. Since the cosmic ray particles follow curved paths, thanks to the vast number of sizable magnetic fields in the Galaxy, they tend to smear themselves out, so to speak, and spread out evenly over the Galaxy regardless of their specific points of origin.

Vast numbers of new cosmic ray particles are constantly being formed by supernovas and, to a lesser extent, from ordinary giant stars, and these are steadily being accelerated and made more energetic. If accelerated sufficiently, they fly out of the Galaxy altogether, and, in addition, vast numbers are constantly striking stars and other objects in the Galaxy. It may be that after 15 billion years of galactic existence, an equilibrium has been reached and as many cosmic ray particles disappear as are formed. For that reason, we can argue that the cosmic ray intensity in Earth's vicinity will remain more or less constant over the eons.

There is, however, one possible exception to this state of affairs. If a supernova should explode in the near vicinity of Earth, there might be trouble. I have discussed such nearby supernovas earlier and pointed out that the chances are very small that one will bother us in the foreseeable future. Even so, I discussed only the light and heat we would receive from such objects. What about the cosmic rays we would receive, since the distance from a nearby supernova to ourselves would then be too short to allow sufficient spreading and blurring by magnetic fields?

In 1968, the American scientists K. D. Terry and W. H. Tucker pointed out that a good large supernova would emit cosmic rays at a rate a trillion times as intense as the sun does and do so for the space of at least a week. If such a supernova were 16 light-years away, the cosmic ray energies reaching us from even that vast distance would be equal to the sun's total radiation over that period of time and that should be sufficient to give every one of us (and of most other forms of life, too, perhaps) enough radiation sickness to kill us. The additional heat supplied by such a supernova and the heat wave that would result would be a matter of no importance at all in that case.

Of course, there are no stars that close to us capable of exploding into a giant supernova and there haven't been any quite that bad in the past, as far as we know, and won't be any in the foreseeable future. However, even a supernova considerably farther away could do considerable damage.

At the present time the intensity of cosmic rays reaching the top of the Earth's atmosphere amounts to about 0.03 rads per year, and it would take 500 times that, or 15 rads per year to do damage. And yet, judging from the frequency of supernovas and from their random positions and sizes, Terry and Tucker calculate that Earth could receive a concentrated dose of 200 rads, thanks to supernova explosions, every 10 million years or so, on the average, and considerably larger doses at correspondingly longer intervals. In the 600 million years since the fossil record began there is a reasonable chance that at least one 25,000 rad flash has reached us. Surely, this could cause trouble—but then there are natural mechanisms that decrease the effectiveness of the cosmic ray bombardment.

For instance, I have just stated that the cosmic ray intensity reaches a certain level at the top of Earth's atmosphere. That was said deliberately, for the atmosphere is not entirely transparent to cosmic rays. As the cosmic rays hurtle past the atoms and molecules making up the atmosphere, sooner or later there are collisions. The atoms and molecules are smashed and particles go flying out of them as "secondary radiation."

The secondary radiation is less energetic than the "primary radia-

tion" of cosmic ray particles in open space, but they are still energetic enough to do plenty of damage. However, they, too, strike
atoms and molecules in Earth's atmosphere and by the time the flying
particles reach the actual surface of the Earth, the atmosphere has
absorbed a substantial portion of the energy.

In short, the atmosphere acts as a protective blanket—not a completely efficient one, but not completely inefficient either. Astronauts
in orbit around the Earth, or on the moon, are subjected to a more
intense bombardment of cosmic rays than we are on Earth's surface
and this is something that has to be taken into account.

Astronauts on relatively short trips beyond the atmosphere may be
able to absorb the additional radiation, but this would not be true of
extended stays in space settlements, for instance. The settlements
would have to be designed with walls thick enough to give at least the
amount of protection against cosmic rays that Earth's atmosphere
does.

Indeed, if the time comes when the major portion of humanity is
ensconced in space settlements and considers itself free of the sun's
viscissitudes and indifferent to the possibilities of the sun's becoming
first a red giant and then a white dwarf, the ebb and flow of the cosmic
ray flux may represent its chief concern and the main possibility of
catastrophe for it.

Of course, getting back to Earth, there's no reason to assume that
the protective action of the atmosphere will ever fail and make us
more subject to the blow of an increase in cosmic ray intensity, at
least not while the atmosphere retains its present structure and composition. There is, however, another kind of protection Earth offers
us that is both more efficient and less durable, and to explain that will
require a little backtracking.

Earth's Magnetic Field

About 600 B.C., the Greek philosopher Thales (624–546 B.C.) first
experimented with naturally magnetic minerals and discovered that
they could attract iron. Eventually, it was learned that the magnetic
mineral lodestone (which we now know to be an iron oxide) could be
used to magnetize thin slivers of steel, which would then display the
properties more intensely than lodestone itself would.

During the Middle Ages, it was discovered that if a magnetic needle
were placed on a light, floating object, that needle would invariably

line up in a north-south direction. One end of the needle was there-fore called the magnetic north pole, the other the magnetic south pole. The Chinese were the first to record this fact some time before 1100, and about a century after that Europeans had picked up the notion.

It was the use of a magnetized needle as a "mariner's compass" that made the European navigators secure at sea and led to the great voyages of discovery beginning soon after 1400, voyages that gave Europe the overlordship of the world for a period of nearly five cen-turies. (The Phoenicians, Vikings, and Polynesians had made remark-able sea voyages without compasses, but only by taking greater risks.)

The ability of a compass needle to find north seemed quite myste-rious at first, and the least mystical explanation supposed that in the far north there was a mountain built of magnetic ore which attracted the needles. Naturally, tales arose of ships venturing dangerously close to this vast magnet. When this happened, the magnet pulled the nails out of the ship which then fell apart and was wrecked. One such tale is found in the *Arabian Nights*.

The English physician William Gilbert (1544–1603) advanced a much more interesting explanation in 1600. He had shaped a piece of lodestone into a sphere and had studied the direction in which a compass needle pointed at various places in the neighborhood of the sphere. He found that it behaved, with reference to the magnetized sphere, just as it behaved with reference to the Earth. He suggested, therefore, that the Earth itself was a huge magnet, with a north mag-netic pole in the Arctic and a south magnetic pole in the Antarctic.

The north magnetic pole was located in 1831 on the western shore of Boothia Peninsula, the northernmost extension of North America, by the Scottish explorer James Clark Ross (1800–62). At that spot, the north-seeking end of the compass needle pointed straight down. The south magnetic pole was located on the rim of Antarctica in 1909 by the Australian geologist Edgeworth David (1858–1934) and the British explorer Douglas Mawson (1882–1958).

But why is the Earth a magnet? Ever since the English scientist Henry Cavendish (1731–1810) had measured the mass of the Earth in 1798, it was known that the average density of the Earth was too high to be made of rock alone. The notion of its center being of metal arose. From the fact that so many meteorites are made of iron and nickel in the proportion of about 10:1 the thought arose that the center of the Earth might be of a similar metal mixture. This was first suggested in 1866 by the French geologist Gabriel August Daubrée (1814–96).

Toward the end of the nineteenth century, the manner in which earthquake waves traveled through the body of the Earth was studied in great detail. It could be shown that those waves which penetrated as deeply as 2,900 kilometers (1,800 miles) under the surface, underwent a sharp change in direction.

It was suggested in 1906 that there was an abrupt change in chemical composition at that point; that the waves had passed from the rocky mantle into the metallic core. This is now accepted. The Earth has a nickel-iron core that is a sphere of about 6,900 kilometers (4,300 miles) in diameter. This core makes up one-sixth the volume of the Earth and, because of its high density, fully one-third of its mass.

It is tempting to suppose that this iron core is a magnet and that this accounts for the behavior of the compass needle. Yet that can't be so. In 1896, the French physicist Pierre Curie (1859–1906) showed that a magnetic substance loses its magnetism if it is heated to a high enough temperature. Iron loses its magnetic properties at a "Curie point" of 760° C. For nickel, the Curie point is 356° C.

Can it be that the nickel-iron core is higher than the Curie point? Yes, for certain types of earthquake waves never pass into the core from the mantle. They are precisely the type that cannot travel through the body of a liquid and it is therefore deduced that the core is hot enough to be made of liquid nickel-iron. Since the melting point of iron is 1535° C under ordinary conditions and should be higher still under the great pressures at the core boundary, that alone should show that the core cannot be a magnet in the sense that a piece of ordinary iron would be.

The presence of a liquid core, however, opened new possibilities. In 1820, the Danish physicist Hans Christian Oersted (1777–1851) had shown that it was possible to produce magnetic effects by means of an electric current ("electromagnetism"). If electricity passes through a wire helix, the result is a magnetic effect very much like that that would originate from an ordinary bar magnet which we could imagine to be placed along the axis of the helix.

With this in mind, the German-American geophysicist Walter Maurice Elsasser (1904–) suggested in 1939 that the rotation of Earth could set up eddies in the liquid core: vast, slow swirls of molten nickel-iron. Atoms are made up of electrically-charged subatomic particles and, because of the particular structure of the iron atom, such swirls in the liquid core might produce the effect of an electric current moving round and round.

Since the swirls would be set up by Earth's rotation from west to east, they would turn west to east, too, and the nickel-iron core would then act like a bar magnet lined up north and south.

The magnetic field of the Earth, however, is not an absolutely steady phenomenon. The magnetic poles shift their position with the years and are, for some reason that we cannot explain, some 1,600 kilometers (1,000 miles) removed from the geographic poles. What's more, the magnetic poles are not at exactly opposite sides of the Earth. A line drawn from the north magnetic pole to the south magnetic pole, passes about 1,100 kilometers (680 miles) to one side of the center of the Earth. In addition, the magnetic field varies in intensity from year to year.

With all this in mind, one might wonder what has happened to the magnetic field in the far past and what might happen in the far future. Fortunately, there is a way of telling—about the past at least.

Among the components of the lava spewed out by volcanic action are various weakly magnetic minerals. The molecules of these minerals have a certain tendency to orient themselves along the magnetic lines of force. While the minerals are in liquid form, this tendency is overcome by the random motion of the molecules in response to the high temperature. As the volcanic rock slowly cools down, however, the random motion of the molecules slows down, too, and eventually, the molecules orient themselves north and south. As the rock solidifies, that orientation locks itself in place. Molecule after molecule does so and, finally, whole crystals exist in which we can detect magnetic poles, the north pole pointing northward and the south pole pointing southward, just as is true of a magnetic compass. (We can identify the north pole of a crystal, or of any magnet, for it is the one that repels the north pole of a compass needle.)

In 1906, a French physicist Bernard Brunhes noticed that some volcanic rock-crystals were magnetized in the direction opposite from the normal. Their magnetic north poles (as identified by a compass needle) were pointing southward. In the years since Brunhes's original discovery, large numbers of volcanic rocks have been studied and it has been found that though in many cases the crystals have the north magnetic poles pointing north in the normal manner, in many other cases the crystals have their north magnetic poles pointing south. Apparently, Earth's magnetic field reverses itself periodically.

By measuring the age of the rocks being studied (by any of a number of well-established methods) it turns out that for the last 700,000 years, the magnetic field has been in its present direction, which we might call "normal." For about 1 million years before that it was in a "reverse" position at almost all times, except for two 100,000-year periods within which it was normal.

On the whole, over the last 76 million years, no fewer than 171 reversals of the magnetic field have been identified. The average length of time between reversals is about 450,000 years, and the two possible alignments, normal and reversed, take up an equal length of time in the long run. The length of time between reversals varies widely, however. The longest measured lapse of time between reversals is 3 million years, the shortest 50,000 years.

How does this reversal take place? Do the Earth's magnetic poles, which are known to wander over the surface of the globe, wander all the way—one managing to wander from the Arctic to the Antarctic, the other vice versa? That does not seem likely. If it were to happen, the poles would have been in equatorial regions for some period of time midway between the reversals. There should in that case be some crystals that are oriented more or less east-west and there are none.

What seems much more likely is that it is the intensity of Earth's magnetic field that varies, increasing and then decreasing. It decreases to zero at times and then begins increasing again, but in the other direction. Eventually, it decreases again to zero and begins increasing in the original direction and so on and so on.

This is similar, in a way, to what happens in the sun's sunspot cycle. The sunspots increase in number, then decrease, then start to increase again with a reversed direction to their magnetic field. Then they decrease and start to increase again with the original direction. Just as the peaks of the sunspots are alternately normal and reversed, so the peaks of Earth's magnetic field are alternately normal and reversed. It's just that the variations of Earth's magnetic field intensity are far less regular than the sunspot cycle of the sun.

What seems likely to bring about the variation in the Earth's magnetic field intensity, and the reversal of its direction, are the variations of the speed and direction of the swirling matter in Earth's liquid core. In other words, the liquid core swirls in one direction, faster and faster, then slower and slower, comes to a brief halt, begins swirling in the other direction faster and faster, then slower and slower, comes to a brief halt, begins in the other direction, and so on.

Why the direction of the swirl should shift, why the speed should change, and why so irregularly, we can't say as yet. We do know, however, the manner in which the Earth's magnetic field affects its bombardment by cosmic rays.

In the 1820s, the English scientist, Michael Faraday (1791–1867) originated the concept of "lines of force." These are imaginary lines moving in a curved path from the north magnetic pole of any object to its south magnetic pole, marking out the path along which the magnetic field was at a constant value.

A magnetized particle can move freely along the lines of force. To cross the lines of force takes energy.

The Earth's magnetic field surrounds the Earth with magnetic lines of force connecting its magnetic poles. Any charged particle traveling from outer space must cross these lines of force to reach Earth's surface and, in so doing, loses energy. If it has only a small amount of energy to begin with, it may lose it all and be unable to cross additional lines of force. In that case, it can only move along a line of force, spiraling it tightly and passing from the Earth's north magnetic pole to its south magnetic pole, and back again, over and over.

This is true for many of the particles in the solar wind, so that there is always a vast number of charged particles traveling along the Earth's magnetic lines of force, setting up what is called the "magnetosphere" far out beyond the atmosphere.

Where the magnetic lines of force come together at the two magnetic poles, the particles follow those lines down toward the Earth's surface and strike the uppermost reaches of the atmosphere. There they collide with atoms and molecules and, in the process, give off energy, and produce the auroras, which are such a beautiful feature of the polar skies at night.

Particles that are particularly energetic can cross all of Earth's magnetic lines of force and hit Earth's surface, but always with less energy than with which they started. Further, they are deflected north and south, and the less energy they have, the farther they are deflected.

Cosmic rays are energetic enough to smash through to Earth's surface, but they are somewhat weakened in this way, and are deflected, too, so that there is a "latitude effect." The cosmic rays reach Earth least intensely in the neighborhood of the equator and become more and more intense the farther north and south one goes from the equator.

Since, as it happens, the density of land life declines as one goes north and south of the tropics (sea life is protected to some extent by thicknesses of water) the net result is that not only are cosmic rays weakened by the magnetic field but they are shifted from regions of much life to regions of little life.

While even the concentrations of cosmic rays at the magnetic poles, where they are most intense, do not seem to interfere with life, it does mean that in the long run, the mutagenic character of cosmic rays on life, generally, is lessened by the existence of Earth's magnetic field.

As the intensity of Earth's magnetic field lessens, this protective effect against cosmic rays weakens. At those periods when Earth's magnetic field is undergoing a reversal, the Earth, for a period of time, has no magnetic field to speak of and the influx of cosmic rays is not deflected at all. The tropic and temperate zones, which carry the main load of land life (including human life) are subject to more cosmic ray intensity at that time than at any other.

What if a supernova happened to blast off nearby during such a period of magnetic field reversal? Its effects would then be considerably greater than they would be if the Earth's magnetic field were quite intense. Could it be that one or more of the great dyings happened at a time when a nearby supernova exploded during a magnetic field reversal?

It is not very likely since nearby supernovas happen only very rarely and magnetic field reversals also happen rarely. To have two very rare phenomena coincide would be much more unlikely than for either to happen alone. Still, the coincidence *might* occur. And, if so, what of the future?

Earth's magnetic field seems to have lost about 15 percent of the strength it had in 1670, when reliable measurements were first being made, and, at the present rate of decrease it will reach zero by A.D. 4000. Even if there is no overall increase in cosmic ray particles through any nearby supernova explosion, the number reaching the chief concentrations of humanity in 4000 will be roughly double what they are now and the genetic load of humanity might increase significantly as a result.

This will probably not be very serious, unless a nearby supernova also explodes, and that cannot be, for the nearest possible supernova by 4000 is Betelgeuse and that is not close enough to be disturbing— even in the absence of a magnetic field.

In the farther future, of course, the coincidence may take place, but neither a nearby supernova nor a magnetic field reversal can possibly catch us by surprise. Both will give ample warning and there should be time to improvise protection against the brief blast of cosmic rays.

This is one potential catastrophe, however, which (to repeat) might affect space settlements more dangerously than the Earth itself.

Part IV

Catastrophes of the Fourth Class

Chapter 12

The Competition of Life

Large Animals

Let us stop to summarize again.

Of the catastrophes of the third class which we have discussed, catastrophes in which the Earth as a whole suffers impairment of its habitability, the only really likely untoward event is a new ice age or, conversely, a melting of the present ice sheets. If either of these happens in the normal course of nature, it will take place very slowly and certainly not for some thousands of years, and it can be either endured or, more likely, controlled.

Humanity might then be able to survive long enough to experience a catastrophe of the second class, one in which the sun undergoes changes which make life on Earth impossible. There, the only likely case is that of the sun becoming a red giant some billions of years hence, and that, while it probably cannot be controlled, can be evaded.

Humanity might then be able to survive long enough to experience a catastrophe of the first class, one in which the universe as a whole becomes uninhabitable. There, the most likely event, in my opinion, is the formation of a new cosmic egg. This, it might appear, can neither be controlled nor evaded, so that it would represent the absolute end of life—but it won't happen for perhaps a trillion years, and who knows what technology will be capable of by then.

And yet we cannot feel ourselves to be safe—not even to the point of surviving to the next ice age—for there are dangers, more immediate dangers, that threaten us even though universe, sun and Earth all remain as smiling and benevolent as they are today.

In other words, we must now consider catastrophes of the fourth class, those which threaten the existence of human life on Earth specifically—even though life generally continues on the planet as before.

But what can possibly bring human life to an end, while life remains in existence, generally?

To begin with, human beings form a single species of organism, and extinction is the common lot of species. At least 90 percent of all the species that have ever lived have become extinct, and of those that survive today, a large fraction are not as numerous or as flourishing as they once were. A good many, in fact, are on the point of extinction.

Extinction can result through changes in environment that ruin those species which, for one reason or another, cannot survive those particular changes. We have discussed some types of environmental change and will discuss more. Extinction can also occur, however, in the direct competition between species and the victory of one species or group of species over another. Thus, through most of the world, placental mammals outsurvived and replaced the less advanced marsupials and monotremes competing for life in the same environment. Only Australia retained a flourishing variety of marsupials and even a couple of monotremes, because it had split off from Asia before the placentals had evolved.

Is there any chance, then, that we may in some fashion be wiped out of existence by some other form of life? We are not the only life forms in the world. There are about 350,000 different species of plants known, and perhaps 900,000 different species of animals. There may be another million or two species that exist and have not yet been discovered. Do any of these other species represent a serious danger to us?

In the early history of the hominids, there were dangers of this sort on every hand. Our hominid ancestors, clothed only in their own skin and with only the various parts of their body as weapons were no match at all for the large predators or even for the large herbivores. The first hominids must have gathered food, rifling the possessions of the inactive plant world, and perhaps occasionally, driven by hunger, they fed on such small animals as they were fortunate enough to catch—rather like chimpanzees today. As for anything of human size or larger, the only recourse of the early hominids was to flee or to hide.

However, even at early stages the hominids were learning to use tools. The hominid hand is well-designed to hold a thigh bone or the branch of a tree, and with these a hominid was not weaponless and could face hooves, claws and fangs with more assurance. As hominids with larger brains evolved, and as they learned to design stone axes and stone-tipped spears, the balance began tipping in favor of them. The stone axe was better than a hoof, the stone-tipped spear better than a fang or claw.

Once *Homo sapiens* appeared, and once these began to hunt in packs, they could (at certain risks, of course) bring down large animals. During the last Ice Age, human beings were quite capable of hunting down mammoths. Indeed, it was human hunting that may have brought the mammoth (and other large animals of the time) to extinction.

What's more, the use of fire gave human beings a weapon and a defense that no other species of living thing could either duplicate or guard against, and behind which human beings were quite secure from predation, since other animals, however large and powerful, carefully and sensibly avoided fire. By the time civilization had begun, the large predators had been, in essence, defeated.

To be sure, individual human beings were still helpless if trapped by a lion, a bear, or other large carnivore, or even by an enraged herbivore such as a water buffalo or wild bull. These, however, were pinpricks, though serious enough for the trapped individual human being.

There was no question, even at the dawn of civilization that human beings, if determined to rid an area of some dangerous animal, could always do so, though it might mean casualties. What's more, human beings, properly armed and determined to kill animals for sport, or to capture them for display, could always do so, though again with possible casualties.

Even today, there are individual defeats, but no one can possibly imagine that human beings as a species are in danger from any of the large animals that now exist, or from all of them put together. Indeed, the situation is quite reversed. Humanity can, with scarcely an effort, drive all the large animals of the world to extinction, and indeed must make a deliberate (and sometimes almost despairing) effort *not* to. The battle having been decided, it is almost as though human beings regret the loss of a worthy enemy.

In ancient times, when victory was already secure, there were dim memories, perhaps, of a time when animals were more dangerous, more threatening, more deadly—and life therefore more suspense-filled and exciting. Naturally, none of the known animals could possibly be pictured as dangerous and threatening to the combined efforts of humanity, so that imaginary animals were conjured up. Some of them were deadly through sheer size. In the Bible one reads of "behemoth," which seems to have been the elephant or the hippopotamus, but which the legend makers expanded to enormous dimensions no animal could truly have. We also read of "leviathan" which may have been inspired by the crocodile or the whale, but which was again expanded into impossibility.

Even giants in human shape are mentioned in the Bible and abound in legends and folklore. Thus, there is Polyphemus, the one-eyed giant Cyclops in the *Odyssey* and the giants who threatened young lads with their "Fee fi fo fum" in the English folk tales.

Failing size, animals are given deadlier powers than they, in fact, possess. The crocodile grows wings and breathes fire, becoming the dreaded dragon. Snakes that can, in actuality, kill by their bite, were graduated to the ability to kill by their breath or even by their glance and became basilisks or cockatrices. The octopus or squid may have given rise to tales of nine-headed Hydras (killed by Hercules) or the many-headed Scylla (to whom Odysseus lost six men) or Medusa, with her hair of living snakes, who turned people to stone when they looked at her (and who was killed by Perseus).

There were combinations of creatures. There were centaurs, with the head and torsos of men affixed to the bodies of horses (inspired, perhaps by ordinary farmers catching their first sight of horsemen). There were sphinxes in which the head and torsos of women were attached to the bodies of lions; griffins which were combinations of eagles and lions; chimeras which were combinations of lions, goats, and snakes. There were more benign creatures: winged horses, one-horned unicorns, and so on.

What they all had in common was that they never existed; and even if they did exist, they could not stand before *Homo sapiens*. Indeed, in the legends, they never did; for the knight always killed the dragon in the end. Even if human giants existed, and were as primitive and nonintelligent as they are always described as being, they would offer no danger to us.

Small Animals

Actually, small mammals can represent greater dangers than do large ones. To be sure, an individual small mammal is less dangerous than an individual large one for obvious reasons. There is less energy at the disposal of the smaller one; it is easier to kill; it can less effectively fight back.

Small mammals, however, do not tend to fight back; they flee. And because they are small they can hide more easily, slither into nooks and crannies in which they cannot be seen and from which they cannot easily be extricated. Unless they are being hunted for food, their very smallness tends to increase their unimportance and the chase is the more easily abandoned.

Then, too, a small mammal does not, generally, make its influence felt as an individual. Small organisms tend to be more shortlived than large ones; but living more quickly means coming to sexual maturity sooner and bearing young sooner. What's more a much smaller investment in energy is required for the production of a small mammal than a large one. In small mammals, the length of pregnancy is shorter and the number of young produced at one time is greater than in large mammals.

Thus, a human being is not sexually mature until thirteen or so; the length of pregnancy is nine months; and one woman in her lifetime would be doing well if she had ten children. If a human couple had ten children and if all these married and had ten children and if all these married and had ten children, then, in three generations, the total number of descendants from the original couple would be 1,110.

A Norway rat, on the other hand, is sexually mature at from eight to twelve weeks of age. It can reproduce from three to five times a year with litters of from four to twelve. Such a rat has a lifespan of only three years, but in that lifespan it can easily produce sixty young. If each of these produced sixty young and each of these sixty young, then in three generations 219,660 rats would have been produced, and that in about nine years.

If such rats continued to multiply unchecked for the human lifespan of seventy years, the total number of rats in the final generation only would be 5,000,000,000,000,000,000,000,000,000,000,000,000,000,000, and these would weigh nearly a million trillion times as much as the Earth.

Naturally, they can't all live, and the fact that very few rats live long enough to reproduce to their full potential is not exactly a waste in the larger scheme of things, for rats are an essential part of the diet of larger creatures.

Nevertheless, this "fecundity," this ability to produce many young very quickly means that the individual rat is virtually a cipher and that the slaughter of rats is virtually without effect. Though nearly every rat is killed in an organized crusade against the animal, those that remain can make up the deficit in numbers with discouraging speed. The smaller the organism, in fact, the less important and effective the individual and the more nearly immortal and potentially dangerous the species.

Furthermore, the presence of fecundity hastens the process of evolution. If in any generation, most rats are affected adversely by a certain poison, or are rendered vulnerable by a certain automatic course of behavior, there are bound to be some that are unusually resistant to the poison as a result of a random and fortunate mutation, or who happen to have a course of behavior that renders them less vulnerable. It is these resistant, less vulnerable rats who tend to survive and have offspring, and these offspring are quite likely to inherit the resistance and comparative invulnerability. In the space of a very brief time, therefore, whatever strategy is used to try to reduce the rat population stops working well.

This makes it appear that the rats are malevolently intelligent, but while they are indeed intelligent for so small an animal, they are not *that* intelligent. It is not the individual we are fighting, but the fecund, evolving species.

In fact, it is quite reasonable to suppose that if there is one characteristic among living things that is most conducive to the survival of the species, and that therefore makes the species most successful, it is fecundity.

We are accustomed to thinking of intelligence as the end toward which evolution is striving, judging this from our own standpoint, but it is still questionable whether intelligence at the expense of fecundity is inevitably victorious in the long run. Human beings have virtually destroyed many of the larger species that are not particularly fecund; they have not even made a dent in the rat population.

Another property of great value for survival is the ability to flourish on a wide range of foods. To be capable of eating one and only one particular article of diet makes it possible for an animal to have a finely tuned digestive system and metabolism. The animal suffers no nutritional problems as long as its specialized food supply is plentiful. Thus the Australian koala, which eats only the leaves of eucalyptus trees is in heaven as long as he is in a eucalyptus tree. A narrow diet puts you at the mercy of circumstance, however. Where eucalyptus trees do not exist, neither do koalas (except, artificially, in zoos). If all eucalyptus trees disappeared, so would all koalas—even in zoos.

On the other hand, an animal with a broad dietary can withstand misfortune. The loss of a delectable item means making do with less delectable ones, but one can survive on that. One reason why the human species has flourished more than have other primate species is that *Homo sapiens* is omnivorous and will eat almost anything, while other primates are mostly herbivorous (the gorilla, for instance, entirely so.)

Unfortunately for us, the rat is also omnivorous and whatever variety of food human beings provide for themselves, the rat will find himself satisfied as well. Wherever the human being goes, therefore, the rat comes along. If we were to ask which mammal most nearly threatens us today, we could not say the lion or the elephant, which we can wipe out to the last individual any time we wish. We have to say the Norway rat.

Still, if rats are more dangerous than lions and if, for that matter, starlings are more dangerous than eagles, the worst that can be said for humanity is that the fight against the small mammals and birds is, at present, a stalemate. They, and other organisms like them, are annoying and frustrating and they can't, without a great deal of trouble, be held in check. There is no real danger, however, of their destroying humanity, unless we are dealt a disabling blow in some other fashion first.

But then there are organisms more dangerous than rats or than any vertebrate. If rats, by their small size and fecundity are hard to beat, what of other organisms still smaller in size and still more fecund? What of the insects?

Of all multicellular organisms, insects are by far the most successful, if you consider them from the standpoint of the number of species. Insects are so short-lived and so fecund that their rate of evolution is simply explosive, and there are now some 700,000 species of insects known as compared with 200,000 species of animals of all other kinds put together.

Furthermore, the list of insect species is not complete or even nearly complete. Some 6,000 to 7,000 new insect species are discovered each year and it is quite likely that there are at least 3 million species of insects in existence all together.

As for the number of individual insects, it is incredible. In and about a single acre of moist soil there may be as many as 4 million insects of hundreds of different species. There may be as many as a billion billion insects living in the world right now; some 250 million insects for each man, woman, and child alive. The total weight of insect life on the planet is greater than the total weight of all other animal life put together.

Almost all the different species of insects are harmless to man. Perhaps 3,000 species at most, out of the possible 3 million, are nuisances. These include the insects which live on us, on our food, or on other things we value—flies, fleas, lice, wasps, hornets, weevils, cockroaches, carpet beetles, termites, and so on.

Some of them are much more than nuisances. In India, for instance, there is an insect called the "red cotton bug" which lives on the cotton plant. Each year, half of all the cotton grown in India is destroyed by them. The boll weevil feeds on the cotton plant in the United States. We can fight the boll weevil more effectively than the Indians can fight the cotton bug. Still, as a result of boll-weevil damage, each pound of cotton produced in the United States costs ten cents more than it would if the boll weevil didn't exist. The losses resulting from insect damage to human crops and human property in the United States alone run to something like eight billion dollars each year.

The traditional weapons developed by human beings in primitive times were aimed at the large animals that the human beings most feared. They grow increasingly ineffective as the target grows smaller. Spears and arrows which are excellent against deer are only of marginal value against rabbits or rats. And to aim a spear or an arrow at a locust or a mosquito is so ridiculous that probably no sane man has ever done it.

The invention of cannons and handguns did nothing to improve the situation. Even nuclear weapons will not kill off the small animals as easily and as thoroughly as it will kill off humanity itself.

To begin with, then, biological enemies were used against the small animals. Dogs, cats, and weasels were used to catch and destroy rats and mice. The small carnivores are better able to follow the rodents wherever they might go; and since those small carnivores are on the quest for food rather than merely to abate a nuisance they are more ardent and single-minded in their pursuit than human beings would be.

Cats, in particular, may have been tamed in Egyptian times not so much for their qualities of companionship (which is about all we expect of them these days) but for their expertise in killing small rodents. By doing so, the cats stood between the Egyptians and the destruction of their grain supply. It was either cats or starve, and it is no wonder the Egyptians deified the cat and made it a capital offense to kill one.

There are biological enemies of insects, too. Birds and the smaller mammals and reptiles are all ready to consume insects. Even some insects consume other insects. Choose the proper predator, the proper time, the proper conditions and you can go a great distance toward controlling a particular insect pest.

The use of such biological warfare was not something, however, that could be handled by early civilizations and the insect-equivalent of the cat was not to be found. In fact, there was no useful method of insect control at all till about a century ago, when poison sprays came into use.

In 1877, compounds of copper, lead and arsenic became the method for fighting the insect enemy. One insect-poison that was much used was "Paris green" which is copper acetoarsenite. This was reasonably effective. Paris green did not affect the plants it was sprayed on. The plants fed on inorganic materials from air and soil, and were powered by energy from sunlight. Traces of mineral crystals on their leaves interfered with none of this. Any insect attempting to eat the leaves, however, would be promptly killed.

Such mineral "insecticides" have their drawbacks.* For one thing, they are poisonous to animal life other than insects—and that means human life as well. Moreover such mineral poisons are persistent. The rain tends to remove some of the mineral and wash it down to the soil. Little by little, the soil accumulates copper, arsenic, and other elements and these eventually reach the roots of plants. In that way they do affect plants adversely and the soil is gradually poisoned.

---

* The term "pesticide" has come into use in recent years, since undesirable organisms other than insects have been attacked with chemicals.

What's more, such minerals can't be used on human beings themselves. They are therefore ineffective against the insects that make humans their prey.

Naturally, there were attempts to find chemicals that would harm only insects and would not accumulate in the soil. In 1935, a Swiss chemist Paul Müller (1889–1965) began to search for such a chemical. He wanted one that could be made cheaply, that had no odor, and that was harmless to noninsect life. He searched among organic compounds—carbon compounds related to those found in living tissue—hoping to find one that would not be as persistent in the soil as mineral compounds were. In September, 1939, Müller came across "dichlorodiphenyltrichloroethane" for which the commonly used abbreviation is DDT. This compound had first been prepared and described in 1874, but for sixty-five years its insecticidal properties had remained unknown.

Many other organic pesticides were discovered and the human war against insects took a strongly favorable turn.

Not entirely, though. The insect powers of evolutionary change remained to be reckoned with. If insecticides killed all but a handful of insects who happened to be relatively resistant to DDT or to other chemicals of the sort, these survivors rapidly multiplied into a new resistant strain. If the same insecticides had killed off insect competitors or predators even more efficiently, the new resistant strain of the insect originally attacked might, for a while, flourish to a greater extent and in greater numbers than before insecticides had been used. To control them, the concentration of insecticide had to be increased and new insecticides brought into use.

As insecticides came to be used more and more widely, more and more indiscriminately, and in greater and greater concentration, other disadvantages showed up. Insecticides might be relatively harmless to other forms of life, but not entirely so. They were often not easily destroyed within the animal body, and animals feeding on insecticide-treated plants stored the chemicals in their fat reserves and passed it on to the other animals who ate them. The stored insecticides could produce damage. As an example, they disturbed the eggshell-forming mechanism in some birds, cutting the birthrate drastically.

The American biologist Rachel Louise Carson (1907–64) published *Silent Spring* in 1962, a book which effectively emphasized the dangers of the indiscriminate use of organic pesticides. New methods have since been developed: pesticides of lesser toxicity; the use of biological enemies; the sterilization of male insects by radioactive radiations; the use of insect hormones to prevent fertilization or maturation of insects.

On the whole, the battle against insects proceeds well enough. There is no sign that human beings are winning in the sense that insect pests will be permanently removed, but we are not losing either. As in the case of rats, the war stands at stalemate, but there is no sign that humanity is going to suffer a disastrous defeat. Unless the human species is badly weakened for other reasons, it is not likely that the insects we are fighting will destroy us.

Infectious Disease

An even greater danger to humanity than the effect of small, fecund pests on human beings, their food, and their possessions, is their tendency to spread some forms of infectious disease.*

Every living organism is subject to disease of various sorts, where disease is defined in its broadest sense as "dis-ease," that is, as any malfunction or alteration of the physiology or biochemistry that interferes with the smooth workings of the organism. In the end, the cumulative effect of malfunctions, misfunctions, nonfunctions, even though much of it is corrected or patched up, produces irreversible damage—we call it old age—and, even with the best care in the world, brings on inevitable death.

There are some individual trees that may live five thousand years, some cold-blooded animals that may live two hundred years, some warm-blooded animals that may live one hundred years, but for each multicellular individual death comes as the end.

This is an essential part of the successful functioning of life. New individuals constantly come into being with new combinations of chromosomes and genes, and with mutated genes, too. These represent new attempts, so to speak, at fitting the organism to the environment. Without the continuing arrival of new organisms that are not mere copies of the old, evolution would come to a halt. Naturally, the new organisms cannot perform their role properly unless the old ones are removed from the scene after they have performed their function of producing the new. In short, the death of the individual is essential to the life of the species.

* As will soon appear, such disease is associated with living organisms still smaller, more fecund, and more dangerous even than insects.

It is essential, however, that the individual not die before the new generation has been produced; at least, not in so many cases as to ensure the population dwindling to extinction.

The human species cannot have the relative immunity to harm from individual death possessed by the small and fecund species. Human beings are comparatively large, long-lived, and slow to reproduce, so that too rapid individual death holds within it the specter of catastrophe. The rapid death of unusually high numbers of human beings through disease can seriously dent the human population. Carried to an extreme, it is not too hard to imagine it wiping out the human species.

Most dangerous in this respect is that class of malfunction referred to as "infectious disease." There are many disorders that affect a particular human being for one reason or another and may kill him or her, too, but which will not, in itself, offer a danger to the species, because it is strictly confined to the suffering individual. Where, however, a disease can, in some way travel from one human being to another, and where its occurrence in a single individual may lead to the death of not that one alone but of millions of others as well, then there is the possibility of catastrophe.

And, indeed, infectious disease has come closer to destroying the human species in historic times than have the depredations of any animals. Although infectious disease, even at its worst, has never yet actually put an end to human beings as a living species (obviously), it can seriously damage a civilization and change the course of history. It has, in fact, done so not once, but many times.

What's more, the situation has perhaps grown worse with the coming of civilization. Civilization has meant the development and growth of cities and the crowding of people into close quarters. Just as fire can spread much more rapidly from tree to tree in a dense forest than in isolated stands, so can infectious disease spread more quickly in crowded quarters than in sparse settlements.

To mention a few notorious cases in history:

In 431 B.C., Athens and its allies went to war with Sparta and its allies. It was a twenty-seven-year war that ruined Athens and, to a considerable extent, all of Greece. Since Sparta controlled the land, the entire Athenian population crowded into the walled city of Athens. There they were safe and could be provisioned by sea, which was controlled by the Athenian navy. Athens would very likely have won a war of attrition before long and Greece might have avoided ruin, but for disease.

In 430 B.C., an infectious plague struck the crowded Athenian population and killed 20 percent of them, including their charismatic leader, Pericles. Athens kept on fighting but it never recovered its population or its strength and in the end it lost.

Plagues very frequently started in eastern and southern Asia, where population was densest, and spread westward. In A.D. 166, when the Roman Empire was at its peak of strength and civilization under the hard-working philosopher-emperor Marcus Aurelius, the Roman armies, fighting on the eastern borders in Asia Minor, began to suffer from an epidemic disease (possibly smallpox). They brought it back with them to other provinces and to Rome itself. At its height, 2,000 people were dying in the city of Rome each day. The population began to decline and did not reach its preplague figure again until the twentieth century. There are a great many reasons advanced for the long, slow decline of Rome that followed the reign of Marcus Aurelius, but the weakening effect of the plague of 166 surely played a part.

Even after the western provinces of the empire were torn away by invasions of the German tribes, and Rome itself was lost, the eastern half of the Roman Empire continued to exist, with its capital at Constantinople. Under the capable emperor Justinian I, who came to the throne in 527, Africa, Italy, and parts of Spain were retaken and, for a while, it looked as though the empire might be reunited. In 541, however, the bubonic plague struck. It was a disease that attacked rats primarily, but one that fleas could spread to human beings by biting first a sick rat and then a healthy human being. Bubonic disease was fast-acting and often quickly fatal. It may even have been accompanied by a more deadly variant, pneumonic plague, which can leap directly from one person to another.

For two years the plague raged, and between one-third and one-half of the population of the city of Constantinople died, together with many people in the countryside outside the city. There was no hope of uniting the empire thereafter and the eastern portion, which came to be known as the Byzantine Empire, continued to decline thereafter (with occasional rallies).

The very worst epidemic in the history of the human species came in the fourteenth century. Sometime in the 1330s, a new variety of bubonic plague, a particularly deadly one, appeared in central Asia. People began to die and the plague spread outward, inexorably, from its original focus.

Eventually, it reached the Black Sea. There on the Crimean peninsula, jutting into the north-central coast of that sea, was a seaport called Kaffa where the Italian city of Genoa had established a trading post. In October, 1347, a Genoese ship just managed to make it back to Genoa from Kaffa. The few men on board who were not dead of the plague were dying. They were carried ashore and thus the plague entered Europe and began to spread rapidly.

Sometimes one caught a mild version of the disease, but often it struck violently. In the latter case, the patient was almost always dead within one to three days after the onset of the first symptoms. Because the extreme stages were marked by hemorrhagic spots that turned dark, the disease was called the "Black Death."

The Black Death spread unchecked. It is estimated to have killed some 25 million people in Europe before it died down and many more than that in Africa and Asia. It may have killed a third of all the human population of the planet, perhaps 60 million people altogether or even more. Never before or after do we know of anything that killed so large a percentage of the population as did the Black Death.

It is no wonder that it inspired abject terror among the populace. Everyone walked in fear. A sudden attack of shivering or giddiness, a mere headache, might mean that death had marked one for its own and that no more than a couple of dozen hours were left in which to die. Whole towns were depopulated, with the first to die lying unburied while the survivors fled to spread the disease. Farms lay untended; domestic animals wandered uncared for. Whole nations— Aragon, for instance, in what is now eastern Spain—were afflicted so badly that they never truly recovered.

Distilled liquors had first been developed in Italy about 1100. Now, two centuries later they grew popular. The theory was that strong drink acted as a preventive against contagion. It didn't, but it made the drinker less concerned which, under the circumstances, was something. Drunkenness set in over Europe and it stayed even after the plague was gone; indeed, it has never left. The plague also upset the feudal economy by cutting down on the labor supply very drastically. This did as much to destroy feudalism as did the invention of gunpowder.*

There have been other great plagues since, though none to match

* Perhaps the most distressing sidelight of the Black Death is the horrible insight into human nature that it offers. England and France were in the early decades of the Hundred Years War at the time. Although the Black Death afflicted both nations and nearly destroyed each, the war continued right on. There was no thought of peace in this greatest of all crises faced by the human species.

the Black Death in unrivaled terror and destruction. In 1664 and 1665 the bubonic plague struck London and killed 75,000.

Cholera, which always simmered just below the surface in India (where it is "endemic") would occasionally explode and spread outward into an "epidemic." Europe was visited by deadly cholera epidemics in 1831 and again in 1848 and 1853. Yellow fever, a tropical disease, would be spread by sailors to more northern seaports, and periodically American cities would be decimated by it. Even as late as 1905, there was a bad yellow fever epidemic in New Orleans.

The most serious epidemic since the Black Death, was one of "Spanish influenza" which struck the world in 1918 and in one year killed 30 million people the world over, and about 600,000 of them in the United States. In comparison, four years of World War I, just preceding 1918, had killed 8 million. However, the influenza epidemic killed less than 2 percent of the world's population, so that the Black Death remains unrivaled.

Infectious disease can strike species other than *Homo sapiens,* of course sometimes with still greater devastation. In 1904, the chestnut trees in the New York Zoological Garden developed the "chestnut blight" and in a couple of decades, virtually every chestnut tree in the United States and Canada was gone. Again, the Dutch elm disease reached New York in 1930 and spread furiously. It is being fought by every resource of modern botanical science, but the elms continue to die and how many can in the end be saved is uncertain.

Sometimes, human beings can make use of animal diseases as a form of pesticide. The rabbit was introduced into Australia in 1859 and, in the absence of natural enemies, it multiplied in wild abandon. Within fifty years it had spread into every part of the continent and nothing that human beings could do seemed to make a dent in its numbers. Then, in the 1950s, a rabbit disease called "infectious myxamatosis," which was endemic among rabbits in South America, was deliberately introduced. It was highly contagious and very deadly to the Australian rabbits who had never been exposed to it before. Almost at once, rabbits were dying by the millions. They were not entirely wiped out, of course, and the survivors are increasingly resistant to the disease, but even now the rabbit population of Australia is well below its peak.

Plant and animal diseases could directly and disastrously affect human economy. In 1872, an epidemic struck the horses of the United States. There was no cure for it. No one at the time understood that it was spread by mosquitoes and before it burned itself out, one-quarter of all American horses were dead. Not only did this represent a serious property loss, but horses were at that time an important source of power. Agriculture and industry were crippled and the epidemic helped bring on a serious depression.

Infectious disease has more than once devastated a harvest and brought disaster. "Late blight" destroyed the potato crop in Ireland in 1845, and one-third of the population of the island starved to death or emigrated. To this day, Ireland has not recovered the population loss of the famine. For that matter, the same disease destroyed half the tomato crop in the eastern United States in 1846.

Infectious disease is clearly more dangerous to human existence than any animal possibly could be, and we might be right to wonder whether it might not produce a final catastrophe before the glaciers ever have a chance to invade again and certainly before the sun begins to inch its way toward red gianthood.

What stands between such a catastrophe and us is the new knowledge we have gained in the last century and a half concerning the causes of infectious disease and methods for fighting it.

Microorganisms

People, throughout most of history, had no defense whatever against infectious disease. Indeed, the very fact of infection was not recognized in ancient and medieval times. When people began dying in droves, the usual theory was that an angry god was taking vengeance for some reason or other. Apollo's arrows were flying, so that one death was not responsible for another; Apollo was responsible for all, equally.

The Bible tells of a number of epidemics and in each case it is the anger of God kindled against sinners, as in 2 Samuel 24. In New Testament times, the theory of demonic possession as an explanation of disease was popular, and both Jesus and others cast out devils. The biblical authority for this has caused the theory to persist to this day, as witness the popularity of such movies as *The Exorcist*.

As long as disease was blamed on divine or demonic influences, something as mundane as contagion was overlooked. Fortunately, the Bible also contains instructions for isolating those with leprosy (a name given not only to leprosy itself, but to other, less serious skin conditions). The biblical practice of isolation was for religious rather than hygienic reasons, for leprosy has a very low infectivity. On biblical authority, lepers were isolated in the Middle Ages, while those with really infectious diseases were not. The practice of isolation, however, caused some physicians to think of it in connection with disease generally. In particular, the ultimate terror of the Black Death helped spread the notion of quarantine, a name which referred originally to isolation for forty (*quarante* in French) days.

The fact that isolation did slow the spread of a disease made it look as though contagion was a factor. The first to deal with this possibility in detail was an Italian physician, Girolamo Fracastoro (1478–1553). In 1546, he suggested that disease could be spread by direct contact of a well person with an ill one or by indirect contact of a well person with infected articles or even through transmission over a distance. He suggested that minute bodies, too small to be seen, passed from an ill person to a well one and that the minute bodies had the power of self-multiplication.

It was a remarkable bit of insight, but Fracastoro had no firm evidence to support his theory. If one is going to accept minute unseen bodies leaping from one body to another and do it on nothing more than faith, one might as well accept unseen demons.

Minute bodies did not, however, remain unseen. Already in Fracastoro's time, the use of lenses to aid vision was well established. By 1608, combinations of lenses were used to magnify distant objects and the telescope came into existence. It didn't take much of a modification to have lenses magnify tiny objects. The Italian physiologist Marcello Malpighi (1628–94) was the first to use a microscope for important work, reporting his observations in the 1650s.

The Dutch microscopist Anton van Leeuwenhoek (1632–1723) laboriously ground small but excellent lenses, which gave him a better view of the world of tiny objects than anyone else in his time had had. In 1677, he placed ditch water at the focus of one of his small lenses and found living organisms too small to see with the naked eye but each one as indisputably alive as a whale or an elephant—or as a human being. These were the one-celled animals we now call "protozoa."

In 1683, van Leeuwenhoek discovered structures still tinier than protozoa. They were at the limit of visibility with even his best lenses, but from his sketches of what he saw, it is clear that he had discovered bacteria, the smallest cellular creatures that exist.

To do any better than van Leeuwenhoek, one had to have distinctly better microscopes and these were slow to be developed. The next microscopist to describe bacteria was the Danish biologist Otto Friedrich Müller (1730–84) who described them in a book on the subject, published posthumously, in 1786.

In hindsight, it seems that one might have guessed that bacteria represented Fracastoro's infectious agents, but there was no evidence of that and even Müller's observations were so borderline that there was no general agreement that bacteria even existed, or that they were alive if they did.

The English optician Joseph Jackson Lister (1786–1869) developed an achromatic microscope in 1830. Until then, the lenses used had refracted light into rainbows so that tiny objects were rimmed in color and could not be seen clearly. Lister combined lenses of different kinds of glass in such a way as to remove the colors.

With the colors gone, tiny objects stood out sharply and in the 1860s, the German botanist Ferdinand Julius Cohn (1828–98) saw and described bacteria with the first really convincing success. It was only with Cohn's work that the science of bacteriology was founded and that there came to be general agreement that bacteria existed.

Meanwhile, even without a clear indication of the existence of Fracastoro's agents, some physicians were discovering methods of reducing infection.

The Hungarian physician Ignaz Philipp Semmelweiss (1818–65) insisted that childbed fever which killed so many mothers in childbirth, was spread by the doctors themselves, since they went from autopsies straight to women in labor. He fought to get the doctors to wash their hands before attending the women, and when he managed to enforce this, in 1847, the incidence of childbed fever dropped precipitously. The insulted doctors, proud of their professional filth, revolted at this, however and finally managed to do their work with dirty hands again. The incidence of childbed fever climbed as rapidly as it had fallen—but that didn't bother the doctors.

The crucial moment came with the work of the French chemist Louis Pasteur (1822–95). Although he was a chemist his work had turned him more and more toward microscopes and microorganisms, and in 1865 he set to work studying a silkworm disease that was destroying France's silk industry. Using his microscope, he discovered a tiny parasite infesting the silkworms and the mulberry leaves that were fed to them. Pasteur's solution was drastic but rational. All infested worms and infested food must be destroyed. A new beginning must be made with healthy worms and the disease would be wiped out. His advice was followed and it worked. The silk industry was saved.

This turned Pasteur's interest to contagious diseases. It seemed to him that if the silkworm disease was the product of microscopic parasites other diseases might be, and thus was born the "germ theory of disease." Fracastoro's invisible infectious agents were microorganisms, often the bacteria that Cohn was just bringing clearly into the light of day.

It now became possible to attack infectious disease rationally, making use of a technique that had been introduced to medicine over half a century before. In 1798, the English physician Edward Jenner (1749–1823) had shown that people inoculated with the mild disease, cowpox, or *vaccinia* in Latin, acquired immunity not only to cowpox itself but also to the related but very virulent and dreaded disease, smallpox. The technique of "vaccination" virtually ended most of the devastation of smallpox.

Unfortunately, no other diseases were found to occur in such convenient pairs, with the mild one conferring immunity from the serious one. Nevertheless, with the notion of the germ theory the technique could be extended in another way.

Pasteur located specific germs associated with specific diseases, then weakened those germs by heating them or in other ways, and used the weakened germs for inoculation. Only a very mild disease was produced but immunity was conferred against the dangerous one. The first disease treated in this way was the deadly anthrax that ravaged herds of domestic animals.

Similar work was pursued even more successfully by the German bacteriologist Robert Koch (1843–1910). Antitoxins designed to neutralize bacterial poisons were also developed.

Meanwhile, the English surgeon Joseph Lister (1827–1912), the son of the inventor of the achromatic microscope, had followed up Semmelweiss's work. Once he learned of Pasteur's research he had a convincing rationale as excuse and began to insist that, before operating, surgeons wash their hands in solutions of chemicals known to kill bacteria. From 1867 on, the practice of "antiseptic surgery" spread quickly.

The germ theory also sped the adoption of rational preventive measures—personal hygiene, such as washing and bathing; careful disposal of wastes; the guarding of the cleanliness of food and water. Leaders in this were the German scientist Max Joseph von Pettenkofer (1818–1901) and Rudolph Virchow (1821–1902). They themselves did not accept the germ theory of disease but their recommendations would not have been followed as readily were it not that others did.

In addition, it was discovered that diseases such as yellow fever and malaria were transmitted by mosquitoes, typhus fever by lice, Rocky Mountain spotted fever by ticks, bubonic plague by fleas and so on. Measures against these small germ-transferring organisms acted to reduce the incidence of the diseases. Men such as the Americans Walter Reed (1851–1902) and Howard Taylor Ricketts (1871–1910) and the Frenchman Charles J. Nicolle (1866–1936) were involved in such discoveries.

The German bacteriologist Paul Ehrlich (1854–1915) pioneered the use of specific chemicals that would kill particular bacteria without killing the human being in which it existed. His most successful discovery came in 1910, when he found an arsenic compound that was active against the bacterium that caused syphilis.

This sort of work culminated in the discovery of the antibacterial effect of sulfanilamide and related compounds, beginning with the work of the German biochemist Gerhard Domagk (1895–1964) in 1935 and of antibiotics, beginning with the work of the French-American microbiologist René Jules Dubos (1901–) in 1939.

As late as 1955 came a victory over poliomyelitis, thanks to a vaccine prepared by the American microbiologist Jonas Edward Salk (1914–).

And yet victory is not total. Right now, the once ravaging disease of smallpox seems to be wiped out. Not one case exists, as far as we know, in the entire world. There are however infectious diseases such as a few found in Africa that are very contagious, virtually 100 percent fatal, and for which no cure exists. Careful hygienic measures have made it possible for such diseases to be studied without their spreading, and no doubt effective countermeasures will be worked out.

New Disease

It would seem, then, that as long as our civilization survives and our medical technology is not shattered there is no longer any danger that infectious disease will produce catastrophe or even anything like the disasters of the Black Death and the Spanish influenza. Yet, old familiar diseases have, within them, the potentiality of arising in new forms.

The human body (and all living organisms) have natural defenses against the invasion of foreign organisms. Antibodies are developed in the blood steam that neutralize toxins or the microorganisms themselves. White cells in the blood stream physically attack bacteria.

Evolutionary processes generally make the fight an even one. Those organisms more efficient at self-protection against microorganisms tend to survive and pass on their efficiency to their offspring. Nevertheless, microorganisms are far smaller even than insects and far more fecund. They evolve much more quickly, with individual microorganisms almost totally unimportant in the scheme of things.

Considering the uncounted numbers of microorganisms of any particular species that are continually multiplying by cell fission, large numbers of mutations must be produced just as continually. Every once in a while such a mutation may act to make a particular disease far more infectious and deadly. Furthermore, it may sufficiently alter the chemical nature of the microorganism so that the antibodies which the host organism is capable of manufacturing are no longer usable. The result is the sudden onslaught of an epidemic. The Black Death was undoubtedly brought about by a mutant strain of the microorganism causing it.

Eventually, though, those human beings who are most susceptible die, and the relatively resistant survive, so that the virulence of the diseases dies down. In that case, is the human victory over the pathogenic microorganism permanent? Might not new strains of germs arise? They might and they do. Every few years a new strain of flu rises to pester us. It is possible, however, to produce vaccines against such a new strain once it makes its appearance. Thus, when a single case of "swine flu" appeared in 1976, a full scale mass-vaccination was set in action. It turned out not to be needed, but it showed what could be done.

Of course, evolution works in the other direction, too. The indiscriminate use of antibiotics tends to kill off those microorganisms that are most successful, while those that are relatively resistant may escape. These multiply and a resistant strain arises which the antibiotic is no longer able to deal with. Thus, we may be creating new diseases, so to speak, in the very act of stemming old ones. There, however, human beings can counterattack by larger doses of old antibiotics or the use of new ones.

It would seem that we can at least hold our own, and that means that we are far ahead of the game, if we consider the situation as it was only two hundred years ago. Yet is it possible that a disease may suddenly strike human beings that is so alien and so deadly that there is no defense possible and that we will be wiped out? In particular, is there the chance that a "plague from space" may arrive, as was postulated in Michael Crichton's best-selling novel, *The Andromeda Strain?*

A prudent NASA takes this into account. They are careful to sterilize objects they send to other planets to minimize the chance of spreading earthly microorganisms to alien soil, thus confusing the possible study of microorganisms native to the planet. And they also place astronauts in quarantine after returning from the moon until they are sure that no lunar infection has struck them.

But this seems an unnecessary caution. Actually, the chances of even microorganismic life elsewhere in the solar system seem small and with every new investigation of the planetary bodies, they seem to grow smaller. Yet what about life from outside the solar system. Here is still another invasion from interstellar space that has not yet been discussed—the arrival of alien forms of microscopic life.

The first to take up this possibility with scientific detachment was the Swedish chemist, Svante August Arrhenius (1859–1927). He was interested in the problem of the origin of life. It seemed to him that life might well be widespread in the universe and that it might spread through infection, so to speak.

In 1908, he pointed out that bacterial spores might be wafted into the upper atmosphere by vagrant winds, and that some might even escape Earth altogether, so that Earth (and any other life-bearing planet, presumably) would leave in its wake a scattering of life-bearing spores. This suggestion is referred to as "panspermia."

Spores, Arrhenius pointed out, could withstand the cold and airlessness of space for very long periods of time. They would be driven away from the sun and out of the solar system by radiation pressure (today, we would say by the solar wind). Eventually, they might arrive at another planet. It was Arrhenius's suggestion that such spores might have arrived at Earth at a time when life had not yet formed on it—that Earth's life was the result of the arrival of such spores and that from those spores we are all descended.*

If this be so, might not panspermia be working today as well? Might not spores be arriving still—at this very moment, perhaps? Might not

---

* In recent years, Francis Crick had pointed out the conceivability of Earth having been deliberately seeded by extraterrestrial intelligences—a kind of "directed panspermia."

some of them be capable of giving rise to disease? Had it been alien spores that had produced the Black Death by any chance? Might they produce another and worse Black Death tomorrow?

One deadly flaw in this line of argument, a flaw that was not apparent in 1908, is that though spores are insensitive to cold and vacuum, they are very sensitive to energetic radiation such as ultraviolet light. They would be likely to be destroyed by radiation from their own star if released by some distant planet, and if they survived that somehow, they would be destroyed by ultraviolet from our own sun even before they approached it closely enough to enter Earth's atmosphere.

Yet might it not be that some spores are relatively resistant to ultraviolet, or are lucky enough to escape? If so, one perhaps need not postulate distant life-bearing planets (for the existence of which there is no direct evidence, although the probabilities in favor seem overwhelming) as the source. What about the clouds of dust and gas that exist in interstellar space and that can now be studied in considerable detail?

In the 1930s, it was recognized that interstellar space contained a very thin sprinkling of individual atoms, predominantly hydrogen, and that interstellar clouds of dust and gas must have a somewhat thicker sprinkling. Astronomers took it for granted, though, that even at their thickest such sprinklings consisted of single atoms. To produce atom combinations, two atoms would have to strike each other and this was not thought to be a very likely event.

Furthermore, if the atom combinations were formed, then, in order to be detected, they would have to be between us and a bright star and absorb some of the light of that star in a characteristic wavelength whose loss we could detect—and be present in such quantity that the absorption would be strong enough to detect. That also seemed unlikely.

In 1937, however, just such requirements were met and a carbon-hydrogen combination (CH, or "methylidene radical") and a carbon-nitrogen combination (CN, or "cyanogen radical") were detected.

After World War II, however, radio astronomy was developed and this became a new and powerful tool for the purpose. In the visible light range, particular atom-combinations could be detected only through their characteristic absorption of starlight. However, individual atoms in such atom-combinations twist, turn, and vibrate, and these motions emit radio waves that could now be detected with great delicacy. Each different atom combination emitted radio waves of

A Choice of Catastrophes   260

characteristic wavelengths, as was known from experiments in the laboratory, and the particular atom combination could then be unmistakably identified. In 1963, no less than four radio wavelengths, all characteristic of the oxygen-hydrogen combination (OH, or "hydroxyl radical") were detected.

Until 1968, only those three two-atom combinations, CH, CN, and OH were known and that was surprising enough. No one expected three-atom combinations to exist, since it would be expecting too much of chance for two atoms to strike each other and cling and for a third atom then to strike.

In 1968, nevertheless, the three-atom molecule of water (H_2O) was detected in interstellar clouds by its characteristic radio-wave radiation, and the four-atom molecule of ammonia (NH_3) as well. Since then, the list of detected chemicals has grown rapidly and combinations of as much as seven atoms have been detected. What's more the more complex combinations all involve the carbon atom, so that it is possible to suspect that even molecules as complex as the amino-acid building-blocks of proteins may exist in space, but in quantities too low to detect.

If we go that far, is it possible that very simple life forms develop in these interstellar clouds? Here we don't even have to worry about ultraviolet light, because stars may be far distant and the dust in the clouds may itself serve as a protective umbrella.

In that case is it further possible that the Earth in passing through such clouds may pick up some of these microorganisms (which the surrounding dust particles may protect even against the ultraviolet radiation of our sun), that they might produce some disease utterly alien to us and against which we would have no defenses, so that we could all die?

The astronomer Fred Hoyle has come closer to home in this respect. He considers the comets, which are known to contain atom combinations much like those in interstellar clouds, in which matter is much more densely packed than in the interstellar clouds; and which, when they approach the sun, release a vast cloud of dust and gas that is driven outward by the solar wind into a long tail.

Comets are much closer to Earth than are interstellar clouds and it is much more likely that the Earth will pass through a comet's tail than through an interstellar cloud. In 1910, as I mentioned earlier in the book, Earth passed through the tail of Halley's comet.

A comet's tail is so thin and vacuumlike that it can't possibly do any gross damage to us such as interfering with Earth's motion or polluting its atmosphere. Could we, however, pick up a few strange microorganisms, which, after multiplying and, perhaps, undergoing mutations in their new environment, would strike with deadly effect? Was the 1918 Spanish influenza born of the passage through the tail of Halley's comet, for instance. Were other vast epidemics produced in this way? If so, might a new passage some time in the future produce a new disease, more deadly than any, and do we face catastrophe at any time, unpredictably, from such an event?

Actually, this all seems in the highest degree unlikely. Even if compounds are formed in interstellar clouds or in comets that are sufficiently complicated to be alive, what are the chances that they would just happen to possess the qualities needed to attack human beings (or any earthly organism).

Remember that only a tiny fraction of all microorganisms are pathogenic and cause disease. Of those that do, most will cause disease in only one particular organism or one small group of organisms and are harmless elsewhere. (For instance, no human being need fear catching Dutch elm disease, and neither need an oak. Nor would either an elm or an oak fear catching cold.)

As it happens, for a microorganism to be effective in causing disease in a particular host, it must be closely and intricately adapted to the task. For an alien organism, formed by chance in the depths of interstellar space or in a comet, to just happen to be chemically and physiologically adapted to successful parasitism on a human being seems quite out of the question.

And yet, the dangers of infectious disease in a new and unexpected form are not entirely done away with even so. There will be occasion to return to the matter and consider it from an altogether different angle later.

Chapter 13

The Conflict
of Intelligence

Nonhuman Intelligence

In the previous chapter, we have discussed the dangers posed to
humanity by other forms of life and have decided that the human
status, against such competing life forms, varies from victory at the
best to stalemate at the worst. And even where stalemate exists,
advancing technology may well bring victory. Certainly, defeat of
humanity by any nonhuman species, while technology remains whole
and civilization is unweakened by other factors, would not seem to
be at all likely.

Those forms of life, however, that we have pictured as without any
real chance of wiping out humanity have this in common—they are
not on the same level of intelligence as is *Homo sapiens*.

Even where nonhuman life wins a partial victory, as where a col-
umn of army ants may overwhelm a person they encounter, or mul-

tiplying plague bacilli may wipe out millions of human beings, the danger is the result of what is more or less automatic and inflexible behavior on the part of the temporarily victorious attackers. Human beings as a species, given a breathing space, can devise a counterattacking strategy and, until now at least, the results of such counterattacks have varied from devastation of the enemy to, at the very least, containment. Nor is the situation likely to grow worse in the future, as nearly as we can tell.

What, however, if the organisms with which we were faced, were as intelligent as ourselves, or even more intelligent? Would we not, in that case, be in danger of being destroyed? Yes, but in all the Earth, where can we find this equal in intelligence?

The most intelligent animals other than human beings—elephants, bears, dogs, even chimpanzees and gorillas—are simply not in our class. None of them can stand up against us for a moment, if humanity were to make use of its technology pitilessly.

If we consider the brain as the physical mediator of intelligence, then the human brain, with its average mass of 1.45 kilograms (3.2 pounds) in the larger of the two sexes, is very nearly the largest that exists either now or in the past. Only those giant mammals, the elephants and the whales, exceed us in this respect.

The largest elephant brain can be as massive as 6 kilograms (13 pounds) or a little over four times that of a human being, while the largest whale brain has the all-time mass record of about 9 kilograms (20 pounds) or over six times that of a human being.

These large brains have far more body to control, however, than the human brain does. The biggest elephant brain may be four times the size of the human brain, but its body is perhaps 100 times that of the human body in mass. Where each kilogram of human brain must handle 50 kilograms of human body, each kilogram of elephant brain must handle 1,200 kilograms of elephant body. In the larger whales, each kilogram of whale brain must handle at least 10,000 kilograms of whale body.

There is less room left over in the elephant brain and in the whale brain for reflection and abstract thought once the needs for coordinating the body are subtracted and there seems to be no question at all that, despite brain size, the human being is far more intelligent than the Asian elephant or the sperm whale.

To be sure, within certain groups of allied organisms, the brain-body ratio tends to increase with decreasing size. In some small monkeys, therefore (and in some humming-birds, for that matter), the ratio is such that each gram of brain need handle only 17.5 grams of body. Here, however, the absolute weights are so small that the monkey brain just isn't large enough to possess the complexity required for reflection and abstract thought.

The human being strikes a happy medium, then. Any creature with a brain much larger than ours has a body so huge that intelligence comparable to ours is impossible. Contrarily, any creature with a brain larger in comparison to its body than is the case for the human being has a brain so small in absolute size that intelligence comparable to ours is impossible.

That leaves us alone at the pinnacle—or almost. Among the whales and their relatives, the brain-body ratio also tends to increase with decreasing size. What about the smallest members of the group. Some dolphins and porpoises are no larger than human beings in weight and yet have brains that are larger than the human brain. The brain of the common dolphin can have a weight of up to 1.7 kilograms (3.7 pounds) and that is 1/6 larger than the human brain. It is more convoluted, too.

Can the dolphin, then, be more intelligent than the human being? Certainly, there seems no question that the dolphin is exceedingly intelligent for an animal. It apparently has complex speech patterns, can learn to put on a good show and clearly has fun doing it. Life in the sea, however, by enforcing streamlining in order to insure rapid motion through a viscous medium has deprived dolphins of manipulative organs equivalent to human hands. Then, too, since the nature of sea water makes fire impossible, dolphins have been deprived of a recognizable technology. For both reasons, dolphins cannot display intelligence in practical human terms.

Dolphins may, of course, possess a deeply introspective and philosophical intelligence, and if we could only understand its system of communication, we might find that its thinking was more admirable than that of human beings. That, however, is irrelevant to the subject matter of this book. Without the equivalent of hands and of technology, dolphins cannot compete with us or endanger us. In fact, human beings, if they put their mind to it (and I hope they never do) could, without undue trouble, wipe out the whale family from top to bottom.

Is it possible, however, that some animals may, in the future, develop intelligence greater than ours and then destroy us? Not likely at all, as long as humanity survives and retains its technology. Evolution does not proceed by vast jumps but at a terribly slow creep. A species will substantially increase its intelligence only over a time interval of a hundred thousand years, or, more likely, a million. There will be ample time for human beings (perhaps growing more intelligent themselves) to notice the change and it seems reasonable to suppose that if humanity conceives a growing danger in the access of intelligence in any species, that that species would be wiped out.* But that brings up another point. Need the intelligent competitor be from Earth itself? Earlier, I spoke of the chances of the arrival of various kinds of objects from the space beyond the solar system—stars, black holes, antimatter, asteroids, clouds of dust and gas, even microorganisms. There remains one more kind of arrival to consider (and the last). What about the arrival of intelligent beings from other worlds? Might not these represent advanced intelligences with a technology far beyond ours? And might they not wipe us out as easily as we could, if we wished, wipe out chimpanzees? Such a thing clearly hasn't happened yet, but might it happen in the future?

This is not something we can completely dismiss. In my book *Extraterrestrial Civilizations* (Crown, 1979), I advance reasons for supposing that technological civilizations may have developed on as many as 390 million planets in our galaxy and that virtually all of them are more advanced technologically than we are. If this were so, the average distance between such civilizations is about 40 light-years. There would thus be an even chance that we are 40 light-years or less from a civilization more advanced than ours. Are we then in danger?

The best reason we might have to feel that we are safe rests in the fact that such an invasion has never taken place in the past, as far as we know, and that for the 4.6-billion-year lifetime of Earth, our planet has been allowed to go its own isolated way. If we have remained untouched for so long in the past, isn't it reasonable to suppose that we will continue to remain untouched for billions of years in the future.

To be sure, there are occasional claims on the part of various irrationalists or quasireligious individuals that extraterrestrial intelligences *have* visited Earth. These frequently gain enthusiastic followers from among those who are not particularly knowledgable in science. There are the tales of the more peculiar "flying-saucer" cultists, for instance, and the claims of Erich von Däniken, whose accounts of "ancient astronauts" have had an enormous allure among the scientifically subliterate.

* There is a special case of the potential speedy access of intelligence in nonhuman terms that does not involve evolution in the ordinary sense of the word. That will be taken up later.

No claim of extraterrestrial invasion either now or in the past has, however, thus far withstood scientific inquiry. Even if cultist claims are allowed, the fact remains that such claimed invasions have proven no danger. Indeed, there are no clear signs that they have affected Earth in any way.

If we cling to rationalism then, we must assume that Earth has always been isolated through all its history and we must ask why. Three general reasons can be offered:

1. There is something wrong with analyses such as those in my book on the subject, and there are, in fact, no civilizations other than ours.

2. If such civilizations do exist, the gap between them is great enough to make crossings impractical.

3. If crossing the gap is practical, and if other civilizations can reach us, they nevertheless choose for some reason to avoid us.

Of these three general suggestions, the first is certainly a possibility and yet most astronomers would doubt that. There is something philosophically repugnant in the thought that of all the stars in the Galaxy (up to three hundred billion of them) only our own sun warms a life-bearing planet. Since there are very many stars like our sun, the formation of a planetary system seems inevitable, the formation of life on any suitable planet also seems inevitable, and the evolution of intelligence and civilization would seem inevitable given enough time.

To be sure, it is conceivable that technological civilizations may develop by the millions but that none survive for very long. The example of our own situation at this moment lends a certain dismal credence to this thought, and yet surely suicide need not be an inevitable consequence. Some of the civilizations should persist. Even ours may.

The third reason also seems doubtful. If crossing the gap between civilizations were possible, then surely expeditions would be sent out to explore and gather knowledge; possibly to colonize. Since the Galaxy is 15 billion years old, there may be at least some civilizations that have lasted a long time and achieved enormously sophisticated levels.

Even if most civilizations are short-lived, those few that are not would colonize the abandoned planets and establish "star empires." And it would seem inevitable, that our solar system would have been reached by the scoutships of such empires and the planets explored.

The flying saucer cultists might well seize upon this line of argument as a rationale for their belief. But if the flying saucers are indeed the scoutships of star empires exploring our planet, why do they not make contact? If they do not wish to interfere with our development, why do they allow themselves to be seen? If they don't care one way or the other about us, why hover about us in such numbers?

Besides which, why have they reached us just now when our technology has been established and never before? Isn't it likely that they would have reached this planet in the interval of billions of years when life was primitive, and might they not have colonized the planet then and established an outpost of their own civilization? There is no sign of such a thing, and pending further evidence, it seems rational to conclude that we have never been visited.

That leaves us with the second reason, which seems the most practical of the three. Even forty light-years is an enormous distance. The speed of light in a vacuum is the maximum velocity at which any particle can travel or any information be transferred. In actual fact, particles with mass always travel at lesser speeds, and objects as massive as spaceships are likely to travel at considerably lesser speeds, even at high levels of technology. (There are, to be sure, speculations concerning the possibility of faster-than-light travel, but they are yet so dim that we have no right to assume they will be realized someday.)

Under these circumstances, it would take several centuries to cross the gap between civilizations even at their nearest, and it doesn't seem likely that vast expeditions of conquest would be sent out.

We might reason that civilizations, once sufficiently advanced, might expand into space, building self-contained and self-sufficient settlements—as someday human beings may. These space settlements may eventually be outfitted with propulsion mechanisms and may set off on voyages through the universe. There may be in the universe such space settlements containing individuals from hundreds or thousands or even millions of different civilizations.

Such wandering settlements, however, might well be as acclimated to space as some forms of life became acclimated to land once they emerged from Earth's ocean. It may be as difficult for organisms on space settlements to land on a planetary surface as it would be for human beings to drop themselves into the abyss. Earth may be occasionally observed from deep space, perhaps, and we might conceive of automated probes being sent into the atmosphere, but very likely nothing more.

On the whole, then, while science fiction has dealt frequently and dramatically with the themes of invasion and conquest by extraterrestrial beings, it is not likely that this offers us any reasonable chance of catastrophe at any time in the foreseeable future.

And, of course, if we continue to survive and if our technological civilization continues to advance, we will become progressively more capable of defending ourselves against outsiders.

War

That leaves humanity with the only intelligent species that it need concern itself as a danger—humanity itself. And that may be enough. If the human species is to be totally eliminated in a Catastrophe of the Fourth Class, it is the human species that may do it.

All species compete among themselves for food, for sex, for security; there are always quarrels and fighting when these needs overlap among individuals. Generally, such quarrels are not to the death, since the individual being worsted generally flees and the victor is generally satisfied with the immediate victory.

Where there is no high level of intelligence, there is no awareness of anything but the present; no clear foresight as to the value of forestalling future competition; no clear memory of past affronts or hurts. Inevitably as intelligence increases, foresight and memory improve, and the point arrives when a victor is not satisfied with the immediate spoils but begins to see the advantage to killing the loser in order to prevent future challenge. Just as inevitably, the point arrives at which a loser who escapes will seek revenge, and if it is clear that a straight individual-to-individual combat will mean another loss, he will seek other means to victory, such as ambush or the gathering of reinforcements.

In short, human beings inevitably reach the level of making war not because our species is more violent and wicked than other species, but because it is more intelligent.

Naturally, as long as human beings were compelled to fight with nails, fists, legs, and teeth alone, little in the way of deadly results could be expected. Bruises and lacerations might be all that would be inflicted in a general way, and the fighting might even be viewed as healthful exercise.

The trouble is that by the time human beings were intelligent enough to plan conflict with the aid of memory and foresight, they had developed the capacity to use tools. As warriors began to swing clubs, wield stone axes, cast stone-tipped spears, and shoot stone-tipped arrows, battles became steadily bloodier. The development of metallurgy made things still worse by substituting for stone, the harder and tougher bronze, and then the still harder and tougher iron. As long as humanity consisted of roving bands of food-collectors and hunters, conflicts would surely have been brief, however, with one side or the other breaking off and fleeing when damage grew unacceptably high. Nor was there any thought of permanent conquest, for ground was not worth conquest. No group of human beings could long maintain themselves in any one place; there was always the necessity of wandering on in search of new and relatively untouched food sources.

A fundamental change came at least as early as 7000 B.C., when the glaciers of the most recent Ice Age were steadily retreating and when human beings were still using stone for tools. At that time in various places in the Middle East (and, eventually, elsewhere as well) human beings were learning to collect food for future use and even to provide for the future creation of food.

They did this by domesticating and caring for herds of animals such as sheep, goats, pigs, cattle, fowl, and making use of them for wool, milk, eggs, and, of course, meat. Properly handled there was no chance of running out, for the animals could be relied on to breed and replace themselves at, if necessary, a greater rate than they were consumed. In this way, food that was inedible or unpalatable to human beings could be used to support animals that were themselves, at least potentially, desirable food.

Even more important was the development of agriculture; the deliberate planting of grain, vegetables, and fruit trees. This made it possible for particular varieties of food to be grown in greater concentration than would exist in nature.

The result of the development of herding and agriculture was the ability of human beings to support a greater density of population than had been possible before. In regions where this advance was made a population explosion took place.

A second result was that society was made static. Herds could not be moved as easily as a human tribe on the prowl could move, but it was agriculture that was crucial here. Farms could not be moved at all. Property and land became important, and the importance of social status resting on the accumulation of possessions increased sharply.

A third result was the increased necessity of cooperation and the development of specialization. A hunting tribe is self-sufficient and the degree of specialization is low. A farming community may be forced to develop and maintain irrigation ditches and to stand guard to keep herds from dispersing or from being carried off by predators (either human or animal). A ditch-digger or a shepherd has little time for other activities, but he can barter his labor for food and other necessities.

Cooperation doesn't necessarily come about through sweet reason, unfortunately, and some activities are harder and less desirable than others. The easiest way of dealing with this problem is for one group of human beings to throw themselves on another and, by killing a few, force the remainder to do all the disagreeable work. Nor can the losers easily flee, tied as they are to farms and herds.

Facing attack by others as an ever present possibility, farmers and herdsmen began to huddle close together and to wall themselves in for protection. The appearance of such walled cities marks the beginning of "civilization"—which comes from a Latin word meaning "city-dweller."

By 3500 B.C., cities had grown to be complex social organizations, containing many people who neither farmed nor herded, but who performed functions necessary for the farmers or herders—whether as professional soldiers, as artisans and artists, or as administrators. By then, the use of metals was coming in and soon after 3000 B.C., writing was developed in the Middle East. This was an organized system of symbols that would record information for longer periods and with less likelihood of distortion than memory alone could. With that the historic period began.

Once cities had developed, each of them in control of a surrounding territory given over to agriculture and herding (the "city-state") wars of conquest became better organized, more deadly and—inevitable.

The early city-states were built up along the course of one river or another. The river offered an easy road of communication for trade and a source of water for the irrigation procedures that made agriculture secure. To have small stretches of the river under the control of separate city-states, always suspicious of each other and usually openly hostile, impaired its use both for communication and irrigation. It was clearly necessary for the common benefit to have the river under the control of a single political unit.

The question was which city-state was to dominate, for the notion of a federal union with all the parts sharing in the decisions never occurred to anyone as far as we know, and would probably not have been a practical course of procedure at that time. The decision of which city-state was to dominate was usually left to the fortunes of war.

The first individual we know of by name who ruled over a considerable stretch of river as a result of a previous history of what may have been military conquest is the Egyptian monarch Narmer (known as Menes in the later Greek accounts). Narmer founded the First Dynasty about 2850 B.C. and ruled over the entire lower Nile Valley. We do not have a circumstantial account of his conquests, though, and his unified rule might possibly have been the result of inheritance or diplomacy.

The first undoubted conqueror, the first man to come to power and then, in a succession of battles, establish his rule over a wide area was Sargon, of the Sumerian city of Agade. He came to power about 2334 B.C. and before his death in 2305 B.C., he had placed himself in control of the entire Tigris-Euphrates valley. Since human beings seem always to have valued and admired the ability to win battles, he is sometimes known as Sargon the Great.

Civilization was well established by 2500 B.C. in four river valleys in Africa and Asia; those of the Nile in Egypt, of the Tigris-Euphrates in Iraq, of the Indus in Pakistan, and of the Huang-Ho in China.

From there, by conquest and by trade, the area of civilization spread outward steadily until, by A.D. 200, it stretched from the Atlantic Ocean to the Pacific in nearly unbroken fashion west to east across the northern and southern shores of the Mediterranean and across southern and eastern Asia. This represents an east-west length of something like 13,000 kilometers (8,000 miles) and a north-south width of from 800 to 1,600 kilometers (500 to 1,000 miles). The total area of civilization may have been, at that time, some 10 million square kilometers (4 million square miles) or about 1/12 the land area of the planet.

What's more, the political units, with time, tended to grow larger, as human beings advanced their technology and became more capable of transporting themselves and material goods, over larger and larger areas. In A.D. 200, the civilized portion of the world was broken up into four major units of approximately equal size.

On the far west, circling the Mediterranean Sea, was the Roman Empire. It reached its maximum physical extent in A.D. 116 and was still virtually intact as late as A.D. 400. East of it and extending over what is now Iraq, Iran, and Afghanistan was the neo-Persian Empire, which, in 226, underwent an accession of strength with the coming to power of Ardashir I, the founder the the Sassanian dynasty. Persia reached its greatest prosperity under Chosroes I about 550 and had a very brief territorial maximum about 620 under Chosroes II.

To the southeast of the Persians was India, which had been nearly united under Asoka about 250 B.C., and was strong again under the Gupta dynasty which came to power about 320. Finally, to the east of India was China, which from about 200 B.C. to A.D. 200 was strong under the Han dynasty.

Barbarians

The ancient wars among the city-states and the empires, which arose out of their conglomeration about some one dominating region, never really threatened catastrophe. There was no question of wiping out the human species since, with the worst will in the world, humanity did not, at that time, possess the power required to do the job.

What was much more likely was that the more or less willful destruction of the painful accumulations of the fruits of civilization might end that aspect of the human adventure. (This would be a catastrophe of the fifth class, something that will occupy the last portion of this book.)

And yet, as long as the quarrel was between one civilized region and another, it was not to be expected that destruction of civilization as a whole would follow—at least, not with the power then in the hands of civilized humanity.

The purpose of war was to extend the power and prosperity of the victor and it suited the conqueror to exact tribute. In order to obtain the tribute enough had to be left to the conquered to enable the tribute to be raised. It was unprofitable to destroy past the point where an object-lesson had been given.

Naturally, where the testimony of the conquered survives, loud are the groans at the cruelty and rapacity of the conqueror, and with justice no doubt—but the conquered survived to groan and, fairly often, survived with enough strength to overthrow the conqueror eventually and become conquerors (just as cruel and rapacious) themselves.

And, on the whole, the area of civilization steadily increased, which is the best indication that the wars, however cruel and unjust to individuals, did not threaten an end to civilization. Indeed, one might argue that the marching armies, as an unintended side effect of their activities, spread civilization; and that stimulus of war-bred emergencies hastened innovation, which sped human technological progress.

There was, however, another kind of warfare that was more dangerous. Every civilized region in ancient times was surrounded by areas of lesser sophistication, and it is customary to refer to the unsophisticated peoples as "barbarians." (The word is of Greek origin, and refers only to the fact that foreigners spoke incomprehensibly with sounds that seemed like "bar-bar-bar" to Greek ears. Even non-Greek civilizations were called "barbarian" by the Greeks. The word has come to be used for uncivilized people, however, with a strong connotation of bestial cruelty.) The barbarians were usually "nomads" (from a Greek word meaning "roaming"). Their possessions were few and consisted chiefly of animal herds, with which they traveled from pasture to pasture as the seasons changed. Their standards of living, by city standards, seemed primitive and poor; and, of course, they lacked the cultural amenities of civilization.

Regions of civilization were, in comparison, wealthy, with their accumulation of food and goods. Those accumulations were a standing temptation to the barbarians, who saw nothing wrong with helping themselves—if they could. Very often, they couldn't. The civilized regions were populous and organized. They had their walled cities for defense and usually understood the science of warfare better. Under strong governments, the barbarians were held at bay.

On the other hand, the people of civilization were pinned to the ground by their possessions and were relatively immobile. The barbarians, on the other hand, were mobile. On their camels or horses, they could raid, and then retreat to raid another day. Victories against them were rarely telling and never (until relatively modern times) final.

Furthermore, many of the civilized population were "unwarlike," for living well, as civilized people do, often leads to the development of a certain lack of toleration for the risky and uncomfortable tasks set soldiers. This means that the greater numbers among the civilized do not count for as much as one might think. A relatively small barbarian warband would find a city population little more than helpless victims, if the civilized army should, for any reason, collapse in defeat.

When a civilized region fell under weak rulers who allowed the army to decay; or when, worse still, the region fell into civil war, a successful barbarian incursion was bound to follow.*

A barbarian takeover was far worse than the routine warfare of civilizations, since the barbarians, unused to the mechanics of civilization, often did not understand the value of keeping the victims alive in order that they might be milked regularly. The impulse was simply to help themselves and to destroy carelessly that which could not at the moment be used. Under such conditions, there would often be a breakdown of civilization over a limited area and for a limited time, at least. There would be a "dark age."

The first example of a barbarian incursion and a dark age followed, naturally enough, not long after our first example of a conqueror. Sargon the Great, his two sons, his grandson and his great-grandson ruled, in succession, over a prosperous Sumero-Akkadian Empire. By 2219 b.c., however, when the great-grandson's rule came to an end, the empire had deteriorated to the point where Gutian barbarians from the northeast were a major problem. By 2180 b.c., the Gutians were in control of the Tigris-Euphrates valley and there followed a century-long dark age.

The barbarians were particularly dangerous when they gained a war-weapon that, temporarily at least, made them irresistible. Thus, about 1750 b.c., the tribes of Central Asia developed the horse-drawn chariot and with that swept down upon the settled lands of the Middle East and Egypt, dominating everything for a period of time.

Fortunately, barbarian invasions have never succeeded in totally wiping out civilization. The dark ages, even at their darkest, were never entirely black, and no barbarians ever failed to feel the attraction of the civilization—even the broken and decaying civilization—of the conquered. The conquerors would become civilized (and, in their turn, unwarlike) and, in the end, civilization would rise again, and usually reach new heights.

There were times when it was a civilized region which gained a new war weapon, and then it might become irresistible in turn. This happened when iron began to be smelted in eastern Asia Minor about 1350 b.c. Gradually, iron become more common, its quality im-

---

* Embarrassed civilized historians sometimes attempt to explain this by speaking of barbarian "hordes." The word "horde" comes from a Turkish word meaning "army" and refers to any loose tribal warband. It has come to carry an impression of great numbers since it seems to excuse defeat at the hands of barbarians, if one can view one's civilized forebears as being overwhelmed by irresistible quantity. Actually, the barbarian "hordes" were almost invariably few in numbers; certainly fewer than those they conquered.

proved, and iron weapons and armor began to be manufactured. When, by A.D. 900, the armies of Assyria became completely "ironized," so to speak, they began a three-century domination of western Asia.

To us of the West, the best-known example of a barbarian invasion and a dark age is that which put an end to the western portion of the Roman Empire. From A.D. 166 onward, the Roman Empire, having passed the expansionist age of its history, fought defensively against barbarian invasion. Time and again, Rome wavered and then regained lost ground under strong emperors. Then, in A.D. 378, the barbarian Goths won a great battle at Adrianople over the Romans, and the Roman legions were forever destroyed. Thereafter, Rome maintained itself for another century by hiring barbarians to fight in its army against other barbarians.

The western provinces came gradually to be under barbarian rule and the amenities of civilization broke down. Italy itself was barbarized, and in 476, the last Roman emperor ruling in Italy, Romulus Augustus, was deposed. A five-century dark age set in, and it was not until the nineteenth century that life in western Europe grew to be as comfortable as it had been under the Romans.

And yet, though we speak of this post-Roman dark age in hushed tones, as though world civilization came within an ace of destruction, it remained a purely local phenomenon, confined to what is now England, France, Germany, and, to some extent, Spain and Italy.

At the low point in 850, when Charlemagne's attempted restoration of some measure of unity and civilization in western Europe had collapsed, and when the region was under the hammer blows of new barbarian raiders—the Northmen from the north, the Magyars from the east—as well as from the civilized Moslems of the south, what was the situation in the rest of the world?

1. The Byzantine Empire, which was the surviving remnant of the eastern half of the Roman Empire, was still strong, and its civilization was preserved in an unbroken line from that of ancient Greece and Rome. What's more, its civilization was actually spreading among the barbarian Slavs, and it was approaching a period of new might under the Macedonian dynasty, a line of warlike emperors.

2. The Abbasid Empire, representing the new religion of Islam, and which had absorbed the Persian Empire and the Syrian and African provinces of the Roman Empire, was at its peak of prosperity and civilization. Its greatest monarch Mamun the Great (son of the famous Harun al-Rashid of the Arabian Nights) had died only in 833.

The independent Moslem realm in Spain was also at a high pitch of civilization (higher in fact than Spain was to see in all the centuries afterward).

3. India, under the Gurjara-Prathihara dynasty, was strong and its civilization continued unbroken.

4. China, though politically unsettled at this time, was at a high point in its culture and civilization, and had successfully spread that civilization to Korea and Japan.

In other words, the total area of civilization was still expanding and only in the far west was there a region that had substantially declined; a region that did not make up more than perhaps 7 percent of the total area of civilization.

Though the barbarian incursions of the fifth century loom so large and fateful in our Western history books, while doing so little damage to civilization as a whole, there were other barbarian incursions in later centuries that were far more threatening. That we are less well acquainted with the later barbarians is only because those regions of western Europe that suffered so badly in the fifth century, suffered less in later centuries.

Throughout the course of history, the steppes of central Asia had bred hardy horsemen who virtually lived on their mounts.* In good years, with sufficient rain, the herds multiplied and so did the nomads. In the years of drought that followed, the nomads led their herds out of the steppes in every direction, thundering against the civilized ramparts from China to Europe.

A succession of tribesmen were, for instance, to be found in what is now the Ukraine in southern Russia, each being replaced by new waves from the east. In the time of the Assyrian Empire, the Cimmerians were to be found north of the Black Sea. They were pushed out by the Scythians about 700 B.C., and these by the Sarmatians about 200 B.C. and these by the Alans about 100 B.C.

About A.D. 300 the Huns approached from the east and they were the most redoubtable of the central-Asia invaders up to that point. In fact, it was their coming that helped push the German barbarians into the Roman Empire. The Germans were not expanding; they were fleeing.

In 451, Attila, the most powerful of the Hunnic monarchs, penetrated as far west as Orleans in France and, near that city, fought a

---

* They were equivalent in some ways to the cowboys of the legendary American West, but where the cowboys flourished for a period of only twenty-five years, the central Asian nomads had been patroling their herds on horseback virtually throughout recorded history.

drawn battle with an allied army of Romans and Germans. That was the farthest west any of the central-Asia tribes was ever to penetrate. Attila died the next year and his empire collapsed almost immediately afterward.

There followed the Avars, the Bulgars, the Magyars, the Khazars, the Patzinaks, the Cumans, with the Cumans still dominating the Ukraine as late as 1200. Each new group of barbarians established kingdoms that looked more impressive on the map than they were in reality, because each consisted of a relatively small population dominating a larger one. Either the small dominating group was shoved aside by another small group from central Asia, or it melted into the dominated group and became civilized—usually both.

Then, in 1162, there was born in central Asia one Temujin. Very slowly, he managed to gain power over first one of the Mongol tribes of central Asia and then another until, in 1206, when he was forty-four years old, he was proclaimed Jenghiz Khan ("very mighty king").

He was now supreme ruler over the Mongols, who, under the new leadership, perfected their style of fighting. Their forte was mobility. On their hardy ponies from which they scarcely ever needed to dismount, they could devour the miles, strike where and when they were not expected, deliver blows too rapidly to be countered, and whirl away before the bewildered foe could mobilize their slow and stupid strength to counterattack.

What had kept the Mongols from making themselves irresistible before then was that they had fought chiefly among themselves and that they had had no leader who knew how to use their potentiality. Under the rule of Jenghiz Khan, however, all civil broils ceased, and in him, they found their military leader. Jenghiz Khan is, in fact, among the greatest captains history records. Only Alexander the Great, Hannibal, Julius Caeser, and Napoleon may fairly be compared with him, and it is quite possible that of them all, he was the greatest. He turned the Mongols into the most remarkable military machine the world had yet seen. The terror of their name grew to the point that the very word of their coming was enough to paralyze those in their path and make resistance impossible.

Before his death in 1227, Jenghiz Khan had conquered the northern half of China and the Khwarezm Empire in what is now Soviet central Asia. More, he had trained his sons and his generals to continue the conquests, which they did. His son, Ogadei Khan, succeeded to the rule, and under him the rest of China was subjugated. Meanwhile, under Batu, a grandson of Jenghiz Khan, and Subutai, the greatest of his generals, the Mongol armies advanced westward.

In 1223, while Jenghiz Khan still lived, a Mongol raid westward had defeated a combined Russian-Cuman army, but that had only been a raid. Now, in 1237, the Mongols poured into Russia. By 1240, they had taken its capital city, Kiev, and virtually all of Russia came under their control. They moved on into Poland and Hungary and, in 1241, defeated a Polish-German army at Liegnitz. They raided into Germany and down to the Adriatic. There was nothing that seemed able to stand against them, and, looking back on it, there seemed no reason to suppose they could not have swept clear to the Atlantic Ocean. What stopped the Mongols was that the word arrived that Ogadei Khan had died and there had to be a vote for a successor. The armies left, and while Russia remained under Mongol rule, the territories west of Russia were free. They had had a bad mauling, but that was all.

In the reigns of Ogadei's successors, Hulagu, another grandson of Jenghiz conquered what is now Iran, Iraq, and eastern Turkey. He took Baghad in 1258. Finally, Kublai Khan (also a grandson of Jenghiz) came to the throne in 1257 and, for a period of thirty-seven years, ruled over a Mongol Empire that included China, Russia, the central Asian steppes, and the Middle East. It was the largest continuous land empire that had ever existed up to that time and of the empires since, only the Russian Empire, and the Soviet Union that followed it, can rival it.

The Mongol Empire had all been built up from nothing by three generations of rulers over a period of half a century.

If ever civilization was shaken from top to bottom by barbarian tribesmen, this was the occasion. (And a hundred years later, there came the Black Death—no worse one-two punch had ever been seen.)

And yet in the end the Mongols, too, did not represent a threat. Their wars of conquest had been bloody and ruthless, to be sure, and were deliberately designed to cow their enemies and victims, for the Mongols were too few in number to be able to rule over such a broad empire unless the inhabitants had been terrified into submission.

It had indeed been in Jenghiz Khan's mind at the start to go further than this (or so it is reported). He played with the thought of destroying the cities and converting the conquered regions into pasture land for nomadic herds.

It is doubtful if he could actually have done this, or that he would not have seen the error of this course of proceeding very soon, even if he had started. As it was, though, he never reached the point of attempting it. Being a military genius, he quickly learned the value of civilized warfare and worked out ways of using the complicated technologies required for laying siege to cities, for scaling and battering walls, and so on. It is but a step from seeing the value of civilization in connection with the arts of war to seeing the value of civilization to the arts of peace as well.

One piece of useless destruction *was* carried out, however. Hulagu's army, having taken the Tigris-Euphrates valley, went on to wantonly destroy the intricate network of irrigation canals that had been spared by all previous conquerors and had kept the area a prosperous center of civilization for 5,000 years. The Tigris-Euphrates valley was turned into the backward and impoverished region it still is today.*

As it happened, though the Mongols became relatively enlightened rulers, not noticeably worse than those who had preceded them, and, in some cases, better. Kublai Khan, in particular was an enlightened and humane ruler under whom vast stretches of Asia experienced a golden age such as they had not had before and were not to have again until (if we stretch a point) the twentieth century. For the first and only time, the vast Eurasian continent came under unitary control from the Baltic Sea to the Persian Gulf, and eastward in a broad path to the Pacific.

When Marco Polo, from the petty patch of land that thought of itself as "Christendom," visited the mighty realm of Cathay, he was awed and thunderstruck, and the people back home refused to believe his descriptions when he wrote them out in all sober truth.

Gunpowder to Nuclear Bombs

It was not long after the Mongol invasions, however, that the see-saw struggle between the citizen farmers and the nomad barbarians swung into an apparently permanent tilt. A military advance came along that gave civilization an edge over the barbarians that the latter could never overtake, so that the Mongols have been called "the last of the barbarians." The invention was gunpowder, a mixture of potassium nitrate, sulfur, and charcoal, which, for the first time, placed an ex-

* The area is in an advantageous position in the last few decades because of the presence of oil under its soil—but that is a temporary resource.

plosive in the hands of humanity.* It required an increasingly elaborate chemical industry to make gunpowder, something barbarian tribes were without.

Gunpowder apparently originated in China, for it seems to have been used there for fireworks as early as 1160. Indeed, it may have been the Mongol invasions, and the clear road their wide empire left for trade, that first brought knowledge of gunpowder to Europe.†

In Europe, however, gunpowder passed from fireworks to a propulsive mechanism. Instead of hurling rocks by catapult, using bent wood or twisted thongs in which to store the propulsive force, gunpowder could be placed in a closed tube (a cannon) with one open end. The cannonball to be hurled would be placed in the open end and the exploding gunpowder would do the propelling.

Very primitive examples of such weapons were used on several occasions in the fourteenth century, most notably at the Battle of Crécy in which the English defeated the French in the opening stages of the Hundred Years War. Such cannon as those used at Crécy were relatively useless, however, and the battle was decided by the English longbowmen whose arrows were far more deadly than the cannon of the day. Indeed, the longbow remained lord of the battlefield (on those occasions when it was used) for another eighty years. It won the Battle of Agincourt for the English in 1415 against a French army far superior in numbers, and a final victory for the English at Verneuil in 1424.

Improvements in gunpowder, however, and improvements in the design and manufacture of cannon, gradually made it possible to get reliable gunpowder artillery, which laid waste the enemy without slaying the gunners themselves. By the latter half of the fifteenth century, gunpowder ruled the battlefield and was to do so for four more centuries.

The French developed artillery, largely to counter the longbow, and the English who had spent eighty years slowly beating France down with those longbows, were driven out again in twenty years by the French artillery. What's more, artillery contributed importantly to the final end of feudalism in western Europe. Not only could the cannonballs beat down the walls of castles and cities without undue

---

* Five centuries before, the Byzantine Empire had disposed of a chemical weapon called "Greek Fire," a mixture of substances (the recipe is not exactly known) which could burn on water. It was used to repel Arab and Russian fleets and several times saved Constantinople from capture. It was not an explosive, however, but an incendiary.
† And of other important technological innovations as well, notably of paper and of the mariner's compass.

trouble, but only a strong central government could afford to construct and maintain an elaborate artillery train, so that little by little the great nobles found themselves forced to knuckle under to the king.

Such artillery meant, once and for all, that the menace of the barbarian was at an end. No horses, however fleet, and no lances, however sure, could stand up against the cannon mouth.

Europe was still in danger from those it was pleased to consider barbarians, but who were as civilized as Europeans were.* The Turks, for instance, had first entered the realm of the Abbasid Empire as barbarians in 840, had helped cause its disintegration (which the Mongols completed) and had survived the Mongol Empire, which had split up into deteriorating fragments soon after the death of Kublai Khan.

In the process, they had become civilized and had captured Asia Minor and sections of the Near East. In 1345, the Osmanli Turks (whose realm came to be known as the Ottoman Empire) crossed into the Balkans and established themselves in Europe—from which they were never to be entirely evicted. In 1453, the Turks captured Constantinople and put a final end to the history of the Roman Empire, but they did so with the help of better artillery than was possessed by any European power.

The conquests of Tamerlane (who claimed descent from Jenghiz Khan) had meanwhile seemed to restore the age of the Mongols and from 1381 to 1405, he won battles in Russia, in the Middle East, and in India. Himself a nomad in spirit, he used the arms and organization of the civilized regions he ruled, and (except for the brief and bloody raid into India) he never moved outside the realms that had previously been conquered by the Mongols.

After the death of Tamerlane, it was at last the turn of Europe. With gunpowder and the mariner's compass, European navigators began to descend upon the shorelines of all the continents, to occupy and populate those that were largely barbarian; to dominate those that were largely civilized. For a period of 550 years, the world became increasingly European. And when European influence began to wane, it was because non-European nations grew more Europeanized, at least in the techniques of warfare, if in nothing else.

---

* Of course, I here use "civilized" only in the sense of possessing cities and a reasonably advanced technology. A nation or people can be civilized in that sense and barbarian in their ruthless lack of humanity. We needn't point to the Turks as an example; the best case in history is that of Germany between 1933 and 1945.

With the Mongols, then, there came a final end to any chance (never very great) of the destruction of civilization through barbarian invasion.

Nevertheless, while civilization was defending itself against barbarism, wars between civilized powers became increasingly savage. Even before the coming of gunpowder, there were cases when civilization seemed in danger of suicide, at least in some areas. In the Second Punic War (218–201 B.C.) the Carthaginian general Hannibal ravaged Italy for sixteen years, and Italy took a long time recovering. The Hundred Years War between England and France (1338–1453) threatened to reduce France to barbarism, and the Thirty Years War (1618–48) finally added gunpowder to the earlier horrors and wiped out half of Germany's population. These wars, however, were restricted in area, and however much Italy or France or Germany might be damaged in this century or that, civilization, as a whole, continued to expand.

But then, as the era of exploration caused European dominion to spread around the world, European wars began to affect outlying continents and the era of world wars began. The first war which might be considered a world war in the sense that armies were engaged on different continents and on the sea—all fighting, one way or another, around issues that were interconnected—was the Seven Years War (1756–63). In this war, Prussia and Great Britain, on one side, fought against Austria, France, Russia, Sweden, and Saxony. The major battles of the war were fought in Germany with Prussia facing impossible odds. Prussia, however, was governed by Frederick II (the Great), the last legitimate monarch to be a military genius, and he was the victor.*

Meanwhile, however, the British and French were fighting in North America, where the war had actually started in 1755. Battles were fought in western Pennsylvania and in Quebec.

Naval battles between Great Britain and France were fought in the Mediterranean, and off both the French coast in Europe and the Indian coast in Asia. Great Britain also fought the Spaniards in the sea off Cuba and the Philippines, while land battles with France were fought in India itself. (Great Britain won, taking Canada from France, and gained an unchallenged foothold in India.)

It was the twentieth century before wars spread at least as far if not

---

* Even his genius could not have won out, however, without British money, and without the fortunate (for him) chance that his inveterate enemy, Empress Elizabeth of Russia, died on January 5, 1762, so that Russia made peace with him.

farther than the Seven Years War did, and at an enormous gain in intensity. World War I saw serious land fighting from France to the Middle East and naval engagements all over the ocean (though the only serious naval battle involving massed warships was fought in the North Sea). World War II saw even more intense action over larger sections of Europe and the Middle East, and over large sections of North Africa and the Far East as well, with naval and air engagements even more widespread and far larger in scale. Nor was it the widening in scale alone that posed a heightening threat to civilization. The advancing level of technology made war weapons steadily more destructive.

The reign of gunpowder came to an end in the late nineteenth century with the invention of high explosives, such as TNT, nitroglycerine, and guncotton. Indeed, the Spanish-American War of 1898 was the last war of any consequence to be fought with gunpowder. Furthermore, ships began to be ironclad and to grow larger; and they carried more powerful guns.

World War I introduced the military use of tanks and airplanes and poison gas. World War II introduced the nuclear bomb. Since World War II, intercontinental ballistic missiles, nerve gases, laser beams, and biological warfare have been developed.

What's more, though war became more extensive, and the weapons of destruction became more powerful, the level of intelligence among generals did not increase. In fact, as the complexities and destructiveness of weapons increased, and as the number of men deployed grew larger, and as the intricacy of combined operations extended over larger areas multiplied enormously, the requirements of quick and intelligent decisions became vastly harder to meet, and generals fell further and further short of the requirements. Generals might not have grown stupider, but they seemed to be stupider relative to the intelligence required.

The American Civil War saw tremendous damage done by incompetent generals, but this sank to insignificance as compared to the damage done by incompetent generals in World War I, and this again decreased by comparison with some of the deadly errors in World War II.

The rule, therefore, that civilized warfare will not destroy civilization, since victors and victims alike are concerned to save the fruits of civilization, no longer applies.

First, the destructiveness of weapons has intensified to such a degree that their full use can not only destroy civilization but even, perhaps, humanity itself; second, the normal incapacity of military leaders to do their job can now lead to mistakes so enormous as to destroy civilization, and even humanity, without that actually being anyone's intention. Finally, at last, we face the one true catastrophe of the fourth class that we can reasonably fear—that an all-out thermonuclear war may somehow start and be carried on, senselessly, to the point of human suicide.

This could happen, but will it?

Let us suppose that the world's political and military leaders are sane and that they retain firm control over the nuclear arsenals. In that case, there is no real chance of nuclear war. Two nuclear bombs have been used in anger—one over Hiroshima, Japan, on August 6, 1945, and one over Nagasaki, Japan, two days later. They were the only two bombs which at that time existed and the intention was to end World War II. In that they succeeded, and there was no possibility of a nuclear counterattack at that time.

For four years, the United States held the only nuclear arsenal but it had no real occasion to use it, since all crises which might provoke war, such as the Soviet blockade of Berlin in 1948, were countered or neutralized without the need to call upon it.

Then, on August 29, 1949, the Soviet Union exploded its first nuclear bomb, and thereafter the possibility of a war with nuclear weapons on each side arose—a war which neither side could win—and a war which both sides well knew that neither side could win.

Attempts to get a sufficiently commanding lead to make war a reasonable possibility failed. Both sides obtained the much more dangerous hydrogen fusion bomb in 1952, both sides developed missiles and satellites, both sides maintained a steady refinement of weaponry in general.

Consequently, war between the superpowers became unthinkable. The most threatening case of a war crisis came in 1962 when the Soviet Union placed missiles on Cuba, ninety miles off the coast of Florida, so that the United States was under the threat of close-range nuclear attack. The United States imposed a naval and air blockade on Cuba and delivered a virtual ultimatum to the Soviet Union to remove those missiles. From October 22 to 28, 1962, the world was as close to nuclear war as it has ever come.

The Soviet Union backed down and removed its missiles. In return, the United States, which had supported an attempt to overthrow Cuba's revolutionary government in 1961, accepted a hands-off policy on Cuba. Each side accepted something of a backdown which would have been unthinkable in prenuclear days.

Again, the United States fought ten years in Vietnam and finally accepted a humiliating defeat, without attempting to use nuclear weapons which would at once have destroyed the enemy. Similarly, China and the Soviet Union did not move toward direct interference in the war, but contented themselves with supporting Vietnam in ways that were far short of war, since they did not want to provoke the United States into a nuclear move.

Finally, in repeated crises in the Middle East in which the United States and the Soviet Union have been ranged on opposite sides, neither of the two superpowers has attempted direct intervention. In fact, the wars of the client-states have not been allowed to continue to the point where one side or the other might be forced to attempt direct intervention.

In short, in the nearly four decades since nuclear weapons have arrived on the scene, they have (except for the proto-explosions over Hiroshima and Nagasaki) never been used in war and the two superpowers have gone to extraordinary lengths to avoid such use.

If this continues, we will not be destroyed by nuclear war—but will it continue? After all, there is nuclear proliferation. In addition to the United States and the Soviet Union, Great Britain, France, China, and India have built nuclear weapons. Others might follow and perhaps inevitably will. Might not a minor power start a nuclear war?

If we assume that the leaders of the minor powers are also sane, then it is hard to see why they should. To have nuclear bombs is one thing; to have a large enough arsenal to avoid swift and sure annihilation from one or the other of the superpowers is quite another. It is likely in fact that any of the minor powers who even makes the faintest gesture toward use of a nuclear bomb will have both superpowers ranged against it at once.

How far can we trust to the assumption of sanity in the world's leaders, however? Nations have, in the past, been under the leadership of psychotic personalities, and even an ordinarily sane leader might, in the grip of rage and despair, be not entirely rational. We can easily imagine someone like Adolf Hitler ordering a nuclear holocaust if the alternative were the destruction of his power, but we might also imagine his underlings refusing to carry out his orders. In point of fact, some of the orders given by Hitler in his final months were *not* carried out by his generals and administrators.

Then, too, there are some national leaders right now who seem to be fanatical enough to push the nuclear trigger if they had one to push. The point is that they don't, and I suspect they are tolerated by the world, generally, precisely because they don't.

Even if all political and military leaders remain sane, is it possible that the nuclear arsenal may get out of their control and that a nuclear war will start through the panicky or psychotic decision of an underling? Worse yet, can it start through a series of small decisions, each one of which seems the only possible response to an enemy move until, finally, the nuclear war starts with no one wanting it and everyone desperately hoping it won't come? (It was in very much this way that World War I started.)

Worst of all, is it possible that world conditions may so deteriorate that a nuclear war may seem an alternative that is preferable to doing nothing?

Undoubtedly, the only certain way to avoid a nuclear war is to destroy all nuclear weapons, and the world may yet come to that before the nuclear war takes place.

Part V

Catastrophes of the Fifth Class

Chapter 14

The Depletion
of Resources

Renewable Items

In the last two chapters, we have decided that the only catastrophe of the fourth class that can possibly befall us is an all-out thermonuclear war sufficiently intense and sufficiently prolonged to destroy all human life—or to leave such inconsiderable remnants of humanity in such miserable condition as to presage eventual extinction.

If this happens, the chances are that other forms of life may also be wiped out but it may well be that insects, vegetation, microorganisms, and so on will survive to repopulate the world eventually and allow it to flourish once again as a habitable planet until such time as (if ever) a new and saner intelligent species evolves.

To be sure, we have argued that the chances are that such an intense and prolonged thermonuclear war will not be resorted to. Even so, lesser degrees of violence would suffice to destroy civilization even if humanity itself were to survive. That would be a catastrophe of the fifth class, the least drastic we shall deal with in this book —but drastic enough.

Suppose, now, that war, together with lesser degrees of violence become things of the past. It is perhaps not very hopeful that this will happen, but it may not be impossible, either. Suppose that humanity decides that war is suicide and makes no sense at all; that it should take the kind of common rational action required to settle disputes short of war, to correct those injustices that breed guerrillism and terrorism, and to then take efficient action to disarm and contain those intransigents whom nothing rational (as defined by the common sense of humanity) will satisfy. Suppose, further, that international cooperation becomes so close as to amount to a form of federalized world government, which can take common action on great problems and great projects.

This may seem hopelessly idealistic, a fairy-tale dream, but suppose it comes to pass. The question, then is: Granted a world of peace and cooperation, are we safe forever? Will we continue to improve our technology until we learn how to prevent the next ice age, 100,000 years from now, and to control Earth's weather to our liking? Will we then continue to further improve our technology as we expand into space and become totally independent of both Earth and sun, so that we can simply move off when it is time for the sun to become a red giant 7 billion years from now (if we have not moved off a long time before)? Will we then continue to further improve our technology until we learn how to survive the contracting universe or the maximizing entropy and outlast even the universe? Or are there dreadful dangers, at close range and nearly, or entirely, inevitable, even in a world at complete peace?

There may be. Consider, for instance, the matter of our improving technology. Throughout this book, I have taken it for granted that technology can and will improve indefinitely if it is given the chance; that it has no natural bounds since knowledge has no limits and can expand indefinitely. But is there no price we must pay for technology; no conditions that must be met? And if so, what happens if we suddenly find that we can no longer pay the price, no longer meet the conditions?

Technology depends for its success on the exploitation of various resources drawn from our environment, and every advance in technology involves an increase in that rate of exploitation, it would seem. In that case, how long can those resources last?

Given the presence of solar radiation for billions of years to come, many of the resources of Earth are indefinitely renewable. Green plants make use of the energy of sunlight to convert water and carbon dioxide to their own tissue substance with oxygen left over in excess and discharged into the atmosphere. Animals ultimately depend on the plant world for food, and combine that food with oxygen to form water and carbon dioxide.

This food and oxygen cycle (to which various minerals essential to life can be added) will continue as long as sunlight does—at least potentially—and, from the human standpoint, both the food we eat and the oxygen we breathe is indefinitely renewable.

Some aspects of the inanimate world are also indefinitely renewable. Fresh water, constantly consumed and constantly running off into the sea, is renewed through evaporation of the oceans by solar heat and by precipitation as rain. Wind will last as long as the Earth is unevenly heated by the sun, the tides will ebb and flow as long as the Earth rotates relative to the moon and sun, and so on.

All forms of life other than human beings deal only with renewable resources. Individual organisms may die through temporary and localized shortages of food or water, or through extremes of temperature, or through the presence and activity of predators, or merely through old age. Whole species may die through genetic change, or through failure to meet minor changes in the environment, or through replacement by other species that are more efficient at survival in one way or another. Life, however, continues because, thanks to the endless cycling of renewable resources, the Earth remains habitable.

Human beings alone deal with nonrenewable resources, and human beings alone therefore run the risk of building a way of life in which something that has become essential may, more or less suddenly, not be there any longer. This disappearance may represent such a dislocation that it may end human civilization. Earth may then remain habitable for life, while no longer suitable for an advanced technology.

The beginnings of technology undoubtedly dealt with renewable resources. The earliest tools had to be those that came ready to hand. A fallen branch of a tree can be used as a club, as can a limb bone of a large animal. These are certainly renewable resources. New branches and new bones we have always with us.

Even when human beings took to throwing stones, no new situation arose. Stones are not renewable in the sense that new stones will not form in a time that is brief compared with that of human activity.

But then, neither are stones consumed by throwing. The thrown stone can eventually be picked up and thrown again. Something new did arrive once stones began to be carefully shaped by chipping, shaving, or grinding so as to create an edge or a point and allow them to be used as knives, axes, spears, or arrowheads.

Here at last is something that is not only nonrenewable but is also consumable. If sharp-edged or sharp-pointed rocks are dulled, they may be rechipped once or twice, but soon enough they are too small to serve their purpose. Generally, new rocks must be sharpened. And though the rocks are always there, large rocks are converted into small ones of which only a fraction are useful. Furthermore, some rocks serve more satisfactorily as sharp-edged tools than others do. Human beings had to start searching for flint, therefore, with some of the avidity that they searched for food.

There was, however, this difference. There was always new food growing, for even the worst droughts and dearths were never permanent. A source of flint, however, once consumed, was consumed for good and would not reappear.

As long as rock was the chief inanimate resource of humanity, there was little fear that it would be utterly consumed. There is too much of it and, at the time that it was the chief inanimate resource (the Stone Age), there were far too few human beings in existence to make an appreciable dent in the supply.

This was true of the use of other varieties of rock—of clay for pottery, of ocher for painting, of marble or limestone for building, of sand for glass, and so on.

The real change came with the use of metals.

Metals

The very word "metal" is from the Greek word meaning "to search for." The metals used for tools and construction these days amount in mass to only about 1/6 of the weight of the rocks making up the crust of the Earth, and almost all of that sixth is not apparent. For the most part the metals exist in combination with silicon and oxygen, or with carbon and oxygen, or with sulfur and oxygen, or with sulfur alone, and form "ores" that are much like other rocks in appearance and properties.

There are a few metals that do not readily form compounds and can exist as nuggets. These are copper, silver, and gold, to which we can add small quantities of meteoric iron. Such free metal is very rare.

Gold makes up only 1/200,000,000 of the crust of the Earth, and is one of the very rarest of the metals, but because it exists almost entirely in the form of nuggets which have a startling and beautiful yellow color, it was probably the first metal to be discovered. It was oddly heavy, shiny enough to serve as an ornament, and soft enough to be beaten into interesting shapes. What's more it was permanent, for it did not rust or in other ways decay.

Human beings may have begun working with gold as long ago as 4500 B.C. Gold, and to a lesser extent silver and copper, were valued because of their beauty and rarity and they became a convenient medium of exchange and an easy way of storing wealth. About 640 B.C., the Lydians of Asia Minor invented coins, small bits of gold-silver alloy of fixed weight, stamped with a government seal to insure their authenticity.

People have generally mistaken the convenience of gold as a medium of exchange for intrinsic value, and nothing has been searched for as ardently or caused such rejoicing when found. Yet gold has no large-scale uses at all. To find a quantity of gold increases the world supply and causes it to lose some of its chief value—rarity.

Consequently, when Spain seized the accumulated gold supplies of the Aztecs and Incas, it did not get rich as a result. The flood of gold into Europe depreciated its value, which meant that prices of all other commodities increased steadily relative to that of gold, and there was an inflation. Spain, which had a weak economy and which had to buy a great many commodities from abroad, found it had to exchange more and more gold for less and less goods.

Nevertheless, the illusion of wealth brought by the gold encouraged Spain to embark on endless wars on the European continent, wars it could not pay for, which drove it into a bankruptcy from which it never recovered—while other nations with developing economies rather than gold grew rich.

The eager attempt during the Middle Ages to find ways of converting other, less valued metals to gold, failed—but the real tragedy would have come if it had succeeded. Gold would quickly have become valueless, and Europe's economy would have been in a turmoil from which it would have been long in recovering.

Other metals, however, which do have intrinsic value, in that they can be used for tools and structures, are, unlike gold, increasingly useful as they become increasingly common. As they become available, and the lower the price relative to gold, the greater the quantity in which they can be used, the stronger the economy, and the higher the standard of living.

For metals to become relatively common, however, human beings had to have more than the nuggets that were occasionally found here and there. Methods had to be discovered for obtaining the metals from their ores; of loosening the metal atoms from combination with the atoms of other elements. This development of "metallurgy" may have taken place as early as 4000 B.C. in the Middle East, with copper the first element to be obtained from its ores.

By about 3000 B.C., it was discovered that certain ores which, as it turned out, contained both copper and arsenic, produced a copper-arsenic alloy that was far harder and tougher than copper alone. This was the first metal that could be used for something more than ornamentation; the first that could be used for tools and armor as an improvement on stone.

Working with arsenic ores is not a safe occupation, however, and arsenic poisoning may have been the first "industrial disease" to plague human beings. Eventually, though, it was discovered that if tin ore was mixed with copper ore, a copper-tin alloy, or "bronze" was obtained which was just as good as the copper-arsenic alloy and far safer to prepare.

By 2000 B.C., the copper-tin variety was in common use and the "Bronze Age" began in the Middle East. The most notable relics we have of that time are Homer's epics *The Iliad* and *The Odyssey* in which warriors fought with bronze armor and bronze-tipped spears.

Copper ore is not common and the civilizations that used bronze intensively found they exhausted their native supplies after a while and had to import quantities from abroad. Tin ore was worse still. Copper is not exactly a common component of the Earth's crust, but tin is still less common. In fact, tin is about 1/15 as common as copper. This meant that by about 2500 B.C., when copper could still be obtained in various places in the Middle East, the local supply of tin seems to have been completely exhausted.

This was the first time in history that human beings had to face the exhaustion of a natural resource; not merely a temporary exhaustion, as of food in time of drought, but a permanent exhaustion. The tin mines were empty and could never be refilled.

Unless human beings were willing to do with only the bronze they had, new supplies of tin would have to be found somewhere. The search continued over a wider and wider area, and by 1000 B.C.,

Phoenician navigators had made their way out of the Mediterranean Sea altogether and had found the "Tin Isles." These, some people think, may have been the Scilly Islands off the southwestern tip of Cornwall.

Meanwhile, though, a technique for obtaining iron from its ores had been developed about 1300 B.C. in Asia Minor. Iron held on more strongly to other atoms than either copper or tin and was more difficult to pry out of combination. It took higher temperatures, and the technique for using charcoal for the purpose took a long time developing.

Meteoric iron was much harder and tougher than bronze, but iron from ores was brittle and fairly useless. The reason for that was that meteoric iron had an admixture of nickel and cobalt. Iron produced from ore, however, was occasionally found to be quite satisfactorily hard and tough. This didn't happen often but it happened often enough to keep metallurgists plugging away at iron-smelting. Eventually, it was discovered that the addition of charcoal to the iron in an appropriate manner hardened it. It produced what we would today call a steely surface.

By 900 B.C., the iron-smelters learned how to do this deliberately and the Iron Age began. It was at once no longer important that copper was rare and that tin was rarer.

This is an example of how human beings have dealt with the exhaustion of resources throughout history. First, they have widened the search for new supplies,* and second, they have found substitutes.

Throughout history, ever since the discovery of metallurgy, the use of metals has been increasing, and at a steadily accelerating rate. New methods for manufacturing steel were discovered in the nineteenth century and metals unknown to the ancients, such as cobalt, nickel, vanadium, niobium, and tungsten were used to mix with steel to form new metal alloys of unexampled hardness or unusual properties. Methods for obtaining aluminum, magnesium, and titanium were developed, and these metals have also come into large-scale use for construction.

But now human beings face the exhaustion of many metals on a worldwide scale, and with it many facets of our technological civili-

---

* A very strong component of the motivation for human exploration rests in the search for resources not available locally. The great voyages of the fifteenth and sixteenth centuries were not primarily intended to increase geographic knowledge or extend European political power. There was a search for products that Europeans lacked, but wanted, such as gold, silk, and spices.

zation. Even old metals have gained new uses we would not easily abandon. Neither copper nor silver is needed for ornamentation, nor even for coinage, but copper has been, till now, essential for our vast electrical network since no substance is as useful as copper for the conducting of electricity, while silver compounds are essential in photography. (Gold, however, remains, to this day, without large-scale uses.)

What do we do then as the metal mines are exhausted, not just in this area or in that, but all over the Earth? It might seem that there would then be no further metals available and that human beings would have no recourse but to abandon so much of their technology that our civilization would collapse, even through the world were at peace and under a humane planetary government.

Some of our important metals are to be exhausted, by some estimates, within a quarter-century. These include platinum, silver, gold, tin, zinc, lead, copper, and tungsten. Does that mean the collapse of civilization is upon us?

Perhaps not. There are ways around such exhaustion.

In the first place, there is conservation. At times, when material is in generous supply, it is used for nonessential purposes, for trivia, for appearance, for fashion. An object made of that material, is replaced when broken, rather than mended or repaired. It may, in fact, be replaced even when it is in perfect working order simply because a new device carries more prestige and social status than an old one would. Deliberate and trivial changes are deliberately introduced sometimes, in order to encourage replacement at a rate faster than required by actual use—simply in order to remain in fashion.

The American economist, Thorstein Veblen (1857–1929) coined the phrase "conspicuous consumption" in 1899 to describe this method of using waste as a sign of social success. Such conspicuous consumption has been part of human social mores from prehistoric times. Until recently, however, it was the prerogative of a thin, aristocratic upper crust, and cast-off items could be used by underlings.

In recent times, however, as the technique of mass-production by machine came into being, it has become possible to spread conspicuous consumption through the population generally. Indeed, waste has at times been considered a necessary means for encouraging production and keeping the economy healthy.

As the supplies of certain commodities dwindle, however, the impulse to conserve will be strengthened in one way or another. Prices will inevitably rise faster than earnings, thus enforcing conservation on those not very well off and restoring the prerogative of waste to the rich alone. If the numerous poor grow sullen and rebellious at the sight of waste in which they cannot participate, society might progress to rationing. This lends itself to abuses, too, but one way or another the dwindling supplies will last longer than one might assume they will, if one judges only by the social mores of prosperity.

Second, substitution: A more common metal can substitute for a less common one. Thus, silver coins have been replaced by those made of nickel and aluminum. Metals, in general, can be replaced by such nonmetals as plastic or glass.

As an example, it is quite possible to use light beams, in place of electrical currents, to transmit messages; and do so, indeed, with very much greater efficiency. Such light beams can be sent along hair-fine threads of glass. Thin cables of glass threads could replace countless tons of copper now being used in electrical communications, and glass, being derived from sand, is not likely to be an easily exhausted resource.

Third, new sources: Though it would seem that all the mines would be exhausted, what we really mean is that all the mines we know about on land would be exhausted. New mines may be discovered, even though this grows increasingly unlikely with time, as more and more of the Earth's surface is minutely examined for ore content.

Then, too, what do we mean by "exhausted"? When we speak of a mine, we speak of a portion of the crust in which a particular metal is sufficiently concentrated so that it may profitably be isolated. With advancing technology, however, methods have been found by which particular metals can be extracted profitably, even when the concentrations are so small that no practical method of exploitation would have existed for them in the past. In other words, mines exist now that would not have been mines in an earlier period.

This process may continue. Although a particular metal may be exhausted if we consider the mines that now exist, new mines may spring up as we find ourselves capable of handling still lower concentrations.

Then, too, we may move off the land altogether. There are sections of the sea bottom that are covered rather thickly with metal nodules. It is estimated that there are 11,000 metric tons of such nodules per square kilometer of ocean floor in the Pacific Ocean. From these nodules, various metals, including some rather useful ones that are in increasingly short supply—such as copper, cobalt, and nickel—can be obtained with scarcely any trouble, once those nodules are dredged up from the sea bottom. Such dredging operations on an experimental basis are now being planned.

Indeed, if the sea bottom, why not the sea itself. Sea water contains every element, usually in very low concentration, since, rain, falling on the land, leaches out a little bit of everything on the way back to the sea. At the moment, we can obtain magnesium and bromine from sea water without undue trouble so that our supplies of those two elements are not likely to run out in the foreseeable future.

After all, the ocean is so huge that the total quantity of any given metal in solution in sea water is surprisingly large, no matter how dilute that solution might be. The sea contains about 3.5 percent of dissolved matter so that each cubic kilometer of sea water contains 36 metric tons of dissolved solids. Another way of putting it is that each metric ton of sea water contains 35 kilograms (77 pounds) of dissolved solids.

Of the dissolved solids in sea water, 3.69 percent is magnesium and 0.19 percent is bromine. A metric ton of sea water would therefore contain 1.29 kilograms (2.84 pounds) of magnesium and 66.5 grams (2.33 ounces) of bromine.* Considering that there are 1,400,000,000,000,000 metric tons of sea water on Earth, one gets an idea of the total quantity of magnesium and bromine available (especially since all that is extracted is eventually slowly washed back into the sea).

A third element, iodine, is also obtained from sea water. Iodine is a comparatively rare element and in a metric ton of sea water only about 50 milligrams (1/600 of an ounce) is present. This is too little to isolate economically by ordinary chemical methods. There are forms of seaweed, however, that can absorb iodine from sea water and incorporate it into their tissues. From the ash of the seaweed, iodine can be obtained.

Might it not be possible to obtain other valuable elements from sea water if techniques were developed to concentrate the often very thin content? The ocean contains, all told, some 15 billion metric tons of aluminum, 4.5 billion metric tons of copper, and 4.5 billion metric tons of uranium. It also contains 320 million metric tons of silver, 6.3 million metric tons of gold, and even 45 metric tons of radium.

They are there. The trick is to get them out.

Or we may move off the Earth, altogether. While not too many years ago, the notion of mining the moon (or even the asteroids)

---

* Neither is present in elementary form, if course, but in the form of dissolved compounds.

would have seemed fit only for science-fiction stories, there are many people who don't find it so terribly impractical now. If the Phoenicians could be driven to the Tin Isles to search for metals in short supply, we can be driven to the moon. The task of mining the moon for us is perhaps no harder than the task of mining the Tin Isles once was for the Phoenicians.

Finally, having gone through the list of new sources, we might even argue that none are really needed. The 81 elements possessing stable atom-varieties are indestructible under ordinary circumstances. Human beings do not consume them, they merely transfer them from one place to another.

Geologic processes, working over billions of years, have concentrated this or that element, including, of course, the various metals, in this region or that. What human beings are doing, and with increasing speed, is to extract the metals and other desired elements from these regions of concentration and to spread them out, more widely, more evenly, more thinly, and mixed them with each other.

The metals are still there, however, though they may be spread out, corroded, and combined with other materials. Indeed, the junkyards of humanity are a vast repository of the various elements he has used, in one form or another, and discarded. Given the proper techniques, they could be recovered and used over again.

Theoretically, then, we cannot run out of the various elements or, in a larger sense, of any substance, since all substances that are not elements are built up of elements.

But mere exhaustion is not the only fate that threatens the resources we use, even the vital resources on which all life, including the human, depends. Even those resources we do not exhaust, and perhaps can never exhaust, may become unusable through our activities. The resources may be there—but they will do us no good.

Pollution

One does not really use up material objects; one merely rearranges atoms. What one uses becomes something else, so that for every consumption there is a balanced production.

If we consume oxygen, we produce carbon dioxide. If we consume food and water, we produce perspiration, urine, and feces. In general, we cannot make use of the products we discard. We cannot profitably breathe carbon dioxide or eat and drink wastes.

Fortunately, the world of life is an ecological unit and what is waste to us is useful material to other organisms. Carbon dioxide is essential to the functioning of green plants and in the process of using it, they produce and discard oxygen. The wastes we produce can be, and are, decomposed and used by a variety of microorganisms and what is left can be used by plants so that in this way water is purified and food is produced. What life discards, life produces again in a vast cycle over and over again. We might call this the "recycling" process.

That is true, to some extent, even in the world of human technology. If human beings burn wood, for instance, they do what lightning does in nature. Human-burned wood enters the cycle along with lightning-burned wood. Throughout hundreds of thousands of years of human use of fire, that use was insignificant compared to lightning fire, so that human activity in no way overloaded the cycle.

Consider, too, the use of stone tools. This involves a steady change of large pieces of rock to small ones. A piece of rock too large for use could be broken into usable portions, and from each usable portion still smaller pieces could be chipped, flaked, or ground off in shaping a tool. Eventually the tool would become useless through the breaking off of smaller pieces that dulled the edge or altered the shape.

This, too, mimics a natural process since the action of wind, water, and temperature-change serve to gradually weather rock into sand. Such small bits of rock can conglomerate again through geologic action. This cycle of large pieces of rock into small and back into large takes place, however, over a very long period of time. By human standards, therefore, the small and useless bits of rock that are the unavoidable waste products of tool manufacture are not recyclable.

Anything produced by human activity that is useless and is not recyclable has, of late, come to be termed "pollution." The small bits of rock were useless, unwanted, and made a mess. As pollution, however, they were relatively benign. They could be easily brushed aside and did no real harm.

Waste products that can be efficiently recycled in nature can nevertheless become pollution if, in a restricted region and time, they overload the capacity of the cycle. When humans burned wood, for instance, they produced ash. This, like the small rocks, could be brushed aside and caused little or no trouble. The burning fire also produced vapors which were largely carbon dioxide and water vapor which, in themselves, cause no trouble. Included in the vapors were minor quantities of other gases which were irritating to the eyes and throat, bits of unburned carbon which smudged surfaces with soot, other finely divided particles which could do damage. The vapors plus these minor constituents made up a visible smoke.

In open air, such smoke quickly dispersed to concentrations too low to be bothersome. There is after all about 5,100,000,000,000,000 metric tons of gases in our atmosphere and the smoke of all the fires of primitive humanity (and of all the forest fires produced by lightning, too) were diluted to insignificance when dispersed in this huge reservoir. Once dispersed, natural processes recycled the substances in the smoke and restored the raw materials that would be used by plants to form wood again.

What, however, if a fire were maintained in a habitation for light, warmth, cooking, and security. Within the habitation, the smoke would accumulate to a high concentration, dirty, smelly, and actively irritating, long before the recycling process could even make a beginning. The result was unbearable, and the smoke of a wood fire was very likely the first example of a pollution problem produced by human technology.

There were several responses that could be made. First, fire could be abandoned altogether, which was probably unthinkable even in the Stone Age. Second, fire could be used only in the open air, which would have caused human beings considerable inconvenience in many ways. Third, a further advance in technology could be used to counter the pollution problem—in short, the equivalent of a chimney (probably a simple hole in the roof, to begin with) could be devised. This third was the solution of choice.

This has been the general way in which human beings, ever since, have dealt with the discomforting side effects. Invariably, the choice has been to move in the direction of additional and corrective technology.

Of course, each bit of corrective technology is quite likely to produce problems of its own and the process may be endless. One can then ask whether the point is reached where an undesirable side effect of technology becomes uncorrectable. Can pollution, for instance, become so extensive that correction will be beyond our reach, and will it then break down our civilization in a catastrophe of the fifth class (or even, perhaps, destroy life in a catastrophe of the fourth class)?

Thus, the old wood fires have increased in number with increasing population. With advancing technology, new fires—of burning fat, coal, oil, and gas—have been added, and the sheer quantity of fire is increasing steadily each year.

Every fire, in one way or another, requires a chimney and the smoke of all of them is discharged into the atmosphere. Right now, this means that about half a billion tons of pollutants in the form of irritating gases and bits of solid are being discharged into the air each year. The atmosphere, as a whole, is in recent decades getting perceptibly dirtier as technology is beginning to overload the cycle.

Naturally, pollution is worst in populated centers, especially industrialized ones, where we now have a "smog" ("smoke" plus "fog") problem. Occasionally, an inversion layer (an upper layer of colder air, trapping a lower layer of warmer air in place for days at a time) prevents dispersal of pollutants and the air becomes dangerous over a limited region. In 1948, there was a "killer smog" over Donora, Pennsylvania, in which twenty-nine people died as a direct consequence. This has also happened on several occasions in London and elsewhere. Even where there are no direct deaths, there is always a long-term increased incidence of pulmonary disease in smoggy areas, up to and including lung cancer.

Is it possible, then, that our technology will leave us with an unbreathable atmosphere in the near future?

The threat is there certainly, but humanity is not helpless. In the early decades of the industrial revolution, cities lay under thick clouds of smoke from burning bituminous coal. A switch to anthracite coal, which produced less smoke, made for a great change for the better in such cities as Birmingham, England, and Pittsburgh, U.S.A.*

Other corrective measures are possible. One danger in smoke rests in the oxides of nitrogen and sulfur that are formed. If nitrogen and sulfur compounds are removed from the fuel to begin with, or if the oxides are precipitated out of the smoke before that smoke is discharged into the atmosphere, many of the fangs of air pollution are drawn. Ideally, the vapors from burning fuel should consist of carbon dioxide and water and nothing more, and it is quite possible that we can attain this ideal.†

---

* There is also a steadily strengthening drive against tobacco addiction, since tobacco smoke includes carcinogens that affect nonsmokers as well as smokers. Unfortunately, tobacco addicts, lost in the grip of their drug, generally ignore or deny this while the tobacco industry would far rather have cancer than lose profits.
† Even the production of carbon dioxide has its dangers, as we shall see.

New varieties of air pollution can turn up unexpectedly. One variety, the potential danger of which was recognized only in the middle 1970s, comes about through the use of chlorofluorocarbons such as Freon. Easily liquefied and entirely nontoxic, they have, ever since the 1930s been used as refrigerants (through alternate vaporization and liquefaction) to replace much more toxic and dangerous gases such as ammonia and sulfur dioxide. In the last couple of decades, they came to be used as a liquid in spray cans. On release in this form, they turn into a vapor and push out, carrying the material they contain in a fine spray.

Though these gases are indeed harmless to life directly, evidence was presented in 1976 to the effect that if they drift into the upper atmosphere, they may diminish and ultimately destroy the ozone layer that exists some 24 kilometers (15 miles) above the Earth's surface. This layer of ozone (an active form of oxygen with molecules made up of three oxygen atoms each, rather than the two each found in molecules of ordinary oxygen gas) is opaque to ultraviolet radiation. It shields the Earth's surface from energetic solar ultraviolet which is dangerous to life. It was perhaps not till the photosynthetic processes of green plants in the sea had produced enough free oxygen to allow an ozone layer to be formed that life was finally able to colonize the land.

If the ozone layer is substantially weakened by the chlorofluorocarbons so that the ultraviolet radiation of the sun reaches Earth's surface with greater intensity, the incidence of skin cancer will increase. Far worse, the effect on soil microorganisms may be drastic and this could violently affect the entire ecological balance in ways we cannot yet foresee but which are very likely to be highly undesirable.

The effect on the ozone layer is still controversial, but already the use of the chlorofluorocarbons in spray cans has greatly decreased and some substitute may have to be found for their use in air conditioners and refrigerators.

Nor is it only the atmosphere that is subject to pollution. There is also Earth's water content or "hydrosphere." Earth's water supply is huge and the mass of the hydrosphere is about 275 times that of the atmosphere. The ocean covers an area of 360 million square kilometers (140 million square miles) or 70 percent of the entire surface area of Earth. The area of the ocean is nearly 40 times the area of the United States.

The average depth of the ocean is 3.7 kilometers (2.3 miles) so that the total volume of the ocean is 1,330,000,000 cubic kilometers (320 million cubic miles.)

Compare this with the needs of humanity. If we consider the use of water for drinking, bathing, washing, and for agricultural and industrial uses, the world uses something like 4,000 cubic kilometers (960 cubic miles) of water per year, only 1/330,000 of the volume of the ocean.

This would sound as though the very concept of a shortage of water were ridiculous, were it not for the fact that the ocean is largely useless to us as a direct water source. The ocean will carry our ships, offer us recreation, and supply us with sea food, but, because of its salt content, we cannot drink it; nor can we use it for washing, for agriculture or for industry. We must have fresh water.

The total supply of fresh water on Earth is equal to 37 million cubic kilometers (8.9 million cubic miles), only 2.7 percent of Earth's total water supply. Most of this is in the form of solid ice in the polar regions and on the mountain peaks and that, too, is not directly useful to us. A good deal of it is in the form of ground water, well below the surface and not easily tapped.

What we need is liquid fresh water on the surface, in the form of lakes, ponds, and rivers, and of that, the supply on Earth is equal to 200,000 cubic kilometers (48,000 cubic miles). This is only about 0.015 percent of Earth's total water supply, but even so it is over 30 times as much fresh water as is used by humanity in a year.

To be sure, humanity doesn't depend on a static supply of fresh water, or we'd use it all up in thirty years at the present rate of consumption. The water we use is recycled naturally. It eventually runs off the land areas into the oceans, while the oceans evaporate in the sun, producing water vapor that eventually falls as rain, sleet, or snow. This precipitation is virtually pure distilled water.

About 500,000 cubic kilometers (120,000 cubic miles) of fresh water precipitates each year. Of this, of course, much falls directly on the oceans and a considerable quantity falls as snow on Earth's ice caps and glaciers. Perhaps 100,000 cubic kilometers (24,000 cubic miles) falls on dry land that is not ice-covered. Even some of this evaporates before it can be used, but about 40,000 cubic kilometers (9,600 cubic miles) are added to the lakes, rivers, and soil of the continents each year (and an equal quantity runs off into the sea). This useful rain supply is still 10 times as much as humanity uses.

Human requirements, however, are rising rapidly. The use of water in the United States has increased tenfold in this century and at this rate it should not be many decades before need will be pressing hard upon supply.

This is the more true in that precipitation is not evenly distributed either in space or in time. There are places where precipitation is in excess and goes to waste, and other places where precipitation is below average and where the population needs every drop that falls. In dry years, there is drought and a sharp decrease in the harvest. The fact is that the usable water supply is dangerously short in many parts of the world right now.

This may be correctable. We might look forward to a time when the weather is controlled and where it can be made to rain on cue in particular areas. The supply of fresh, liquid water can be increased by the direct distillation of sea water—something now being employed in the Middle East—or possibly by freezing the salt out of sea water.

Then, too, the ice supply of the world is restored to the ocean chiefly in the form of icebergs breaking off the edges of the Greenland and Antarctica ice sheets. These icebergs are huge reservoirs of fresh water that melt into the ocean unused. They might, however, be towed to arid coasts and used there.

Then, too, the ground water which underlies even deserts can be tapped more efficiently, and the surfaces of lakes and reservoirs can be coated with thin films of harmless chemicals to cut down evaporation.

The matter of the supply of fresh, liquid water, may not prove a serious problem, therefore. More dangerous is the matter of pollution.

The waste products of all water creatures on Earth are, as a matter of course, deposited in the water in which they live. These wastes are diluted and recycled by natural processes. The waste products of land animals are deposited on the land, where they are decomposed by microorganisms for the most part and recycled. Human wastes follow the same cycle, and these, too, can be recycled although the great concentrations of human population tend to overload the regions in and surrounding large cities.

Worse than this, the chemicals that industrialized humanity uses and produces are discharged into the rivers and lakes and eventually reach the ocean. Thus, in the past century, human beings have begun to use chemical fertilizers containing phosphates and nitrates in great and increasing quantities. They are deposited on land, of course, but the rain washes some of these chemicals into nearby lakes. Since phosphates and nitrates are necessary for life, the growth of organisms in such lakes is greatly encouraged and the process is called "eutrophication" (from Greek words meaning "good growth").

This sounds good, but the organisms that are chiefly encouraged are algae and other one-celled organisms, which grow at tremendous rates and crowd out other forms of life. When algae die, they are decomposed by bacteria which, in the process, consume much of the dissolved oxygen in the lakes so that the lower regions become virtually lifeless. The lake thus loses much of its value as a source of fish or, for that matter, of drinking water. Eutrophication accelerates those natural changes which cause a lake to fill in and turn first into a marsh and then dry land. What would normally happen over the course of thousands of years might conceivably take place in decades.

If this is what will happen in the case of substances useful to life, what about outright poisons?

In many chemical industries, chemicals are produced that are poisonous to life, and wastes containing them are discharged into rivers or lakes. There, it might be thought, they are diluted to harmlessness and destroyed by natural processes. The trouble is that some chemicals exert deleterious effects even after great dilution and are not easily destroyed by natural processes.

Even if the chemicals are not directly harmful at great dilution, they can accumulate in life forms, as the simple forms absorb the poison and the more complex forms eat the simple forms. In that case, even if the water remains drinkable, the water life becomes inedible. And by now, in the industrialized United States, almost every lake and river is polluted to some degree—many badly so.

Of course, all these chemical wastes are eventually washed into the ocean. It might be thought that the ocean, which is so vast, can absorb any quantity of waste products, however undesirable they may be, but this is not so.

In this century, the ocean has had to absorb incredible quantities of petroleum products and other wastes. Through the wreckage of oil tankers, the washing-out of oil tanks, the disposal of automobile waste oil, 2 million to 5 million metric tons of oil find their way into the ocean each year. Ship litter of various kinds amounts to 3 million metric tons per year. Over 50 million metric tons of sewage and other waste enter the ocean each year from the United States alone. Not all of this is dangerous, but some of it is, and the amount of all this material entering the ocean is increasing steadily each year.

The regions near the continental shores, which are most richly endowed with life, are most seriously affected by pollution. Thus, a tenth of the area of coastal waters off the United States that have in the past served as a source of shellfish are now unusable due to pollution.

Water pollution, therefore, if it continues indefinitely, not only threatens usefulness of our essential fresh-water supply in the not-too-distant future, but also the viability of the ocean. If we were to imagine an ocean so poisoned as to become lifeless, we would lose the microscopic green plants ("plankton") that float on or near its surface and that account for 80 percent of the oxygen-renewal of the atmosphere. It is almost certain that life on land could not long survive the death of the ocean.

In short, water pollution might, in the extreme, virtually destroy life on Earth and produce a catastrophe of the fourth class.

Yet this does not have to be. Before those wastes that are dangerous are deposited in the water, they might be treated in such ways as to reduce their deleterious effects; particular poisons might be outlawed altogether and not produced, or destroyed once produced. If water eutrophication takes place, algae may be harvested out of the lake water to remove the excess nitrates and phosphates—which may be used as fertilizers on land once again.

And speaking of land, there are solid wastes, too; wastes that do not enter either the atmosphere or the hydrosphere—garbage, rubbish, litter. These have been produced by human beings from the beginnings of civilization. The ancient cities of the Middle East allowed their garbage and litter to accumulate and eventually built new houses over it. Every ruined ancient city is on its own mound of garbage, and archeologists dig into the garbage to learn from it about the life of those times.

In modern times, we cart the solid wastes away and dump them in unused areas. Every city, therefore, has its areas where countless dead automobiles sit rusting and its mountainous heaps of garbage that serve as the happy hunting grounds of a myriad of rats.

These wastes accumulate without end and large cities, with endless tons of garbage to move each day (more than one ton per person per year on the average in industrialized areas), are running out of places in which to build their mountains.

A serious aspect of the problem is that an increasing percentage of the solid wastes are not easily recyclable by natural processes. Aluminum and plastics are particularly long-lived. And yet ways of recycling them can be developed; in fact, must be developed. It is precisely these dumps, as I indicated earlier, that form a kind of mine of used metals.

Energy: Old

The problems of depletion of resources and pollution of the environment have, then, the same solution—recycling.* Resources are what are withdrawn from the environment, and pollution is what is returned to the environment in excess of what can be safely recycled by natural processes. Human beings must accelerate the recycling process in order to restore resources as rapidly as they are consumed, and to remove pollution as rapidly as it is produced. The cycle must be made to move more quickly and, in some cases, in directions that do not occur in nature.

This requires time, labor, and the development of new and better recycling techniques. It takes one thing more; it takes energy. It takes energy to mine the sea bottom, or to get out to the moon, or to concentrate thin dispersions of elements, or to build up complicated substances from simple ones. It takes energy to destroy undesired wastes, or to treat them into harmlessness, or to collect them, or to retrieve them. No matter how determinedly, how cleverly, and how innovatively we learn to turn the cycle in order to keep the resources coming and the pollution disappearing, it will take energy.

And unlike material resources, energy cannot be used and reused indefinitely; it is not recyclable. While energy cannot be destroyed, the portion of any fixed amount of it that can be converted into work declines steadily in accordance with the second law of thermodynamics. For that reason we have more cause to worry about energy than other resources.

In short, when speaking of the possibility of the exhaustion of recources generally, it would appear we need only consider the possibility of the exhaustion of our energy supply. If we have a plentiful and continuing supply of energy, then we can use it to recycle our material resources and we will exhaust nothing. If we have only skimpy supplies of energy, or if a plentiful supply becomes exhausted, then we lose the ability to manipulate our environment and we lose all other resources as well.

---

* Only material pollution has been discussed here. There are other forms of pollution that cannot be recycled and that will be taken up later.

Where, then, do we stand on energy?

The major source of energy here on Earth is the radiation of the sun, which bathes us constantly. Plant life converts the energy of sunlight into the chemical energy stored in their tissues. Animals, by eating the plants, build up their own stores of chemical energy.

Sunlight is converted into inanimate forms of energy, too. Through the uneven heating of the Earth, currents are set up in the ocean and in the air, and such energy can sometimes be violently concentrated as in hurricanes and tornadoes. By the evaporation of ocean water and its condensation as rain, the energy of flowing water on land is produced.

To a lesser degree, there are nonsolar sources of energy. There is the internal heat of the Earth, which makes itself felt more or less benignly, in the form of hot springs and geysers, and violently, in the form of earthquakes and volcanoes. There is the energy of Earth's rotation which makes itself felt in the tides. There is the energy of radiation from other sources than the sun (stars, cosmic rays) and the natural radioactivity of elements such as uranium and thorium in the soil.

For the most part plants and animals make use of the stores of chemical energy in their tissues, though even simple forms of life can make use of inanimate energy as well—as when plants allow pollen or seeds to be blown by the wind, for instance.

This was true for early human beings, too. They made use of their own muscular energy, transferring it and concentrating it by means of tools. This is not, in itself, to be dismissed lightly. A great deal can be done by means of wheels, levers, and wedges even with only human muscles behind them. The pyramids of Egypt were built in this fashion.

Even before the dawn of civilization, human beings had learned to use the muscles of other animals to eke out their own labor. This represented a gain over the use of human slaves in a number of ways. Animals were more tractable than humans, and animals could eat food that human beings would not so that they did not represent a drain on the food supply. Finally, some animals offer greater concentrations of energy, which they can expend at a faster rate than human beings can.

Perhaps the most successful domesticated animal from the standpoint of speed and power was the horse. Until the beginning of the nineteenth century, human beings could not travel faster overland than a horse could gallop, and the entire agricultural economy of a nation like the United States depended on the number and health of its horses.

Human beings also used inanimate energy sources. Goods could be transported downstream on rafts making use of a river current. Sails could catch the wind which would then drive a ship against the current. Water currents could also be used to turn a water wheel, and the wind could be used to turn a windmill. In ocean ports, ships could make use of the tides to set out to sea.

All these energy sources were limited, however. They either disposed of only a certain amount of power, as a horse did, or they were subject to uncontrollable fluctuation, as was true of the wind, or they were tied down to specific geographic locations as a rapid-current river was.

One turning point came, however, when, for the first time, human beings made use of an inanimate source of energy that was available in any reasonable quantity and for any reasonable time, that was portable, and that was completely controllable—fire.

Where fire is concerned, no other organisms but hominids have ever made the slightest advance in the direction of the use of fire. That is the sharpest dividing line between hominids and all other organisms (I say "hominids" because fire was not first used by *Homo sapiens*. There is definite evidence of fire having been used in caves in China in which the earlier hominid species, *Homo erectus,* dwelt at least half a million years ago).

Fire comes naturally into being when lightning strikes trees, and undoubtedly the first use of fire was only of the preexisting phenomenon. Small bits of lightning-begun fire were salvaged, fed with wood, and not allowed to go out. A lost campfire was an inconvenience since some other fire had to be found to serve as an igniter, and if one could not be found, the inconvenience became a disaster.

It was not till 7000 B.C., probably, that methods for starting a fire from scratch were discovered. How this came about and where and when the method was first used is not known and may never be known, but at least we know that the discovery was made by *Homo sapiens,* for by then (and long before then) it was the only hominid in existence.

The chief fuel for fire in ancient and medieval times was wood.* Like other energy sources, wood was indefinitely renewable—but with a difference. Other energy sources cannot be used any faster

---

* Fats, oils, and waxes, obtained from animals or plants, were used in lamps and candles, but these represented minor contributions.

than they can be renewed. Men and horses tire and must rest. Wind and water have a fixed amount of energy and more cannot be withdrawn. Not so in the case of wood. Plant life continually grows, of course, and replaces itself so that, up to a certain limit, depredations can be made good. Wood can be used at a rate outstripping the rate of renewal so that human beings, in effect, draw against future supplies.

As the use of fire grew steadily greater with the rise of human population and with the development of a more advanced technology, the forests began to disappear in the immediate neighborhood of human centers of civilization.

Nor was it possible to conserve, for virtually every advance in technology raised the requirement for energy and human beings were never willing to abandon their technological advances. Thus, smelting of copper and tin required heat and that meant wood-burning.

The smelting of iron required still more heat, and wood could not produce a high enough temperature. If, however, wood was burned under conditions that allowed little or no air circulation, the center of the woodpile was charred black into almost pure carbon ("charcoal"). This charcoal burned more slowly than wood, produced virtually no light but a much higher temperature than burning wood did. Charcoal made iron smelting practical (and supplied the carbon that produced a steely surface and made the iron useful). Producing the charcoal was, however, very wasteful of the wood itself.

The forests have continued to retreat before the onslaught of civilization but even so they have not entirely vanished. About 10 billion acres of Earth's land area, or some 30 percent of the whole is still forested.

Nowadays, of course, efforts are made to preserve the forest and to use no more than can be replaced. Every year, 1 percent of the growing timber can be harvested and this yields some 2 billion cubic meters of wood. Of this, nearly half is still used as fuel chiefly in the less-developed nations of the world. Probably more wood is burned as fuel today than was burned in earlier times when wood was almost the only fuel, but when world population was far less than it is today. The forests that remain are preserved as well as they are (which is not perfectly well, by the way) only because wood is no longer the chief fuel and energy source of humanity.

A great deal of wood formed in earlier ages of Earth's history did not decay completely, but fell into bogs under conditions that removed other atoms but left carbon behind. This carbon was buried under sedimentary rock and compressed. Large quantities of it exist underground and represent a kind of fossilized wood that is now known as "coal." It represents a chemical store of energy produced by sunlight over a period of some hundreds of millions of years.

It is estimated that in the world today there is something like 8 trillion metric tons of coal distributed over many areas. If this is so, the carbon content of Earth's supply of coal is twice as great as that in the Earth's supply of presently living organisms.

Coal seems to have been burned in China in medieval times. Marco Polo, who visited the court of Kublai Khan in the thirteenth century, reported that black stones were burned as fuel, and it is only after that that it began to be burned occasionally here and there in Europe, first in the Netherlands.

It was in England, though, that the use of coal on a large scale began. Within the narrow confines of that kingdom, the shrinking of the forests had grown serious. Not only was it becoming difficult to satisfy the need to warm homes in England's far-from-sunny climate with home-grown timber and to fulfil also the need to fuel its growing industries, but there were the requirements of England's navy, on which the nation depended for its security.

Fortunately for England, easily obtainable coal was located in the northern section of the land. In fact, there were more surface out-croppings of coal in England than in almost any other region of comparable size. By 1660, England was producing 2 million tons of coal each year, over 80 percent of all the coal produced in the world at that time, and it became a major influence in sparing the increasingly scarce and valuable forests. (Nowadays, coal production in Great Britain is something like 150 million tons per year, but this is only 5 percent of the world production.)

Coal would be particularly useful if it could be used for iron smelting, for the need for charcoal was so wasteful of wood that iron smelting was the chief agency for the destruction of the forests.

In 1603, Hugh Platt (1552–1608) first discovered how to heat coal in order to drive off remaining noncarbonaceous material and to leave behind virtually pure carbon in the form of what was called "coke." Coke proved an admirable substitute for charcoal in iron smelting.

Once the process of coke-making was perfected in 1709 by the English ironmaster Abraham Darby (1678–1717), coal began to take its true place as the primary energy source of the world. It was coal that powered the Industrial Revolution in England, for it was burning coal that heated the water that formed the steam that ran the steam engines that turned the wheels of factories and locomotives and steamships. It was the coal of the Ruhr Basin, of Appalachia, and of the Donetz Basin that made possible the industrialization of Germany, the United States, and the Soviet Union, respectively.

Wood and coal are solid fuels, but there are liquid and gaseous fuels, too. Plant oils could be used as liquid fuels in lamps, and wood, when heated, gave off inflammable vapors. It is, in fact, the combination of these vapors with air that gives rise to the dancing flames of fire. Solid fuels that do not produce vapors, as charcoal and coke for instance, simply glow.

It was only in the eighteenth century, however, that inflammable vapors could be produced and stored. In 1766, the English chemist Henry Cavendish (1731–1810) isolated and studied hydrogen, which he called "fire gas" because of its inflammability. Hydrogen, in burning, develops a great deal of heat, 250 calories per gram, as compared with 62 calories per gram for the best grade of coal.

The disadvantage of hydrogen is that it burns too readily and, if mixed with air before ignition, it explodes with shattering force, if a spark is introduced. Accidental admixture is all too possible.

If ordinary grades of coal are heated in the absence of air, however, inflammable vapors are given off ("coal gas") which is only half hydrogen. The other half contains hydrocarbons and carbon monoxide and the mixture as a whole will burn but is less likely to explode.

The Scottish inventor William Murdock (1754–1839) used jets of burning coal gas to illuminate his house in 1800, proving that the danger of explosion was low. In 1803, he used gas lighting in his factory and by 1807, London streets were beginning to be lit by gas.

Meanwhile, seeping from the rocks, was an inflammable oily material which was eventually called "petroleum" (from the Latin word for "rock oil") or, more commonly, simply "oil." As coal is the product of bygone ages of forests, so oil is the product of bygone ages of past unicellular sea-life.

The more nearly solid outcroppings of such materials were known as "bitumen" or "pitch" to the ancients and were used for waterproofing purposes. The Arabs and Persians did note the inflammability of the liquid portions.

In the nineteenth century, the search was on for gases, or easily vaporized liquids, to feed the demand for lighting, improvements on the then used coal gas and whale oil. Petroleum was a possible source; it could be distilled, and a liquid portion, "kerosene," was ideal for lamps. What was needed was a larger supply of petroleum.

In Titusville, Pennsylvania, there were seepages of petroleum which were collected and sold as patent medicines. A railway conductor, Edwin Laurentine Drake (1819–80), reasoned there was a large supply of petroleum underground and undertook to drill for it. In 1859, he was successful in producing the first productive "oil well," whereupon drilling started elsewhere and the modern petroleum industry was born.

More petroleum has been extracted from the Earth with each year since then. The coming of the automobile and the internal-combustion engine, which run on "gasoline" (a liquid fraction of petroleum that is still more easily vaporized than kerosene), gave the industry an enormous boost. There are also gaseous fractions of petroleum, consisting chiefly of methane (with molecules made up of one carbon atom and four hydrogen atoms) that are called "natural gas."

As the twentieth century opened, oil was beginning to gain appreciably on coal, and after World War II, it became the chief fuel of industry the world over. Thus, where coal supplied 80 percent of Europe's energy needs before World War II, it supplied only 25 percent of those needs in the 1970s. World consumption of oil has more than quadrupled since World War II, and now stands at about 60 million barrels a day.

The total amount of petroleum mined in the world since Drake's first oil well is about 350 billion barrels, with half of that having been used in the last twenty years. The total estimated supply still left in the ground is about 660 billion barrels and at current rates of use that will last only thirty-three years.

This is a serious problem. Oil is the most convenient fuel available in large quantity that human beings have ever been able to find. It is easy to obtain, easy to transport, easy to refine, easy to use—and not only for energy, but as a source for a vast variety of synthetic organic material such as dyes, drugs, fiber, and plastics. It is thanks to oil that industrialization has been spreading throughout the world at an enormous rate.

To switch from oil to any other energy source will require tremendous inconvenience and capital outlay—yet it will surely have to be done eventually. As it is, the steadily rising rate of use, and the prospect of an inevitable fall in production, has been sending the price of oil skyward in the 1970s, and that has been distorting the world economy to a disturbing degree. By 1990, oil production is likely to fall behind demand, and if other energy sources do not take up the slack, the world will face an energy shortage. All the dangers of resource depletion and of air and water pollution will then be sharpened, even as an energy dearth in the home, in the factories, on the farms, raises the problem of the unavailability of heat, of commodities, even of food.

It seems inappropriate, therefore, to fear catastrophes to the universe, to the sun, to the Earth; we need not dread black holes and extraterrestial invasions. Instead, we must ask ourselves whether, within the space of this generation, the supply of available energy, which has been rising steadily all through human history, will finally peak and begin to decline, and whether that will carry down with it human civilization, bring on a desperate last-ditch nuclear war over the waning scraps, and so end all hope of human recovery.

This is the catastrophe that we face more imminently than any other I have discussed.

Energy: New

Although the prospect of energy starvation can be viewed as both imminent and horrible, it is not inevitable. It is a catastrophe that is of human making and is therefore amenable to human postponement or avoidance.

As is the case with depletion of other resources, however, there are possible counterattacks.

First, there is conservation. For two hundred years, humanity has been fortunate enough to have had available to it cheap energy, and that has had its less fortunate side effects. There has been little reason to drive toward conservation and strong temptation to move toward conspicuous consumption.

But the era of cheap energy is over (at least for a time). The United States, for instance, is no longer self-sufficient in oil. It has produced more oil by far than any other nation has, but for that very reason its reserves are now shrinking rapidly even as the national rate of consumption continues to move upward.

It means that the United States must import more and more oil from abroad. This tips the trade balance into a more and more unfavorable direction, places unbearable pressure on the dollar, drives inflation onward, and, in general, steadily erodes the American economic position.

Conservation is, therefore, not only desirable for us, but is absolutely necessary.

There is considerable room for conservation of energy, beginning with the elimination of the greatest energy-waster of all—the various military machines of the world. Since war is impossible without suicide, the maintenance of competing military machines at an astronomic cost in energy, when the world's chief energy resource is shrinking rapidly, is clearly insane.

In addition to direct conservation of oil, there are clear possibilities for an increase in the efficiency with which oil can be withdrawn from already existing oil wells, so that dry wells can, to some extent, be made to flow again.

There can also be an increase in the efficiency with which energy can be extracted from burning oil (or burning fuel generally). At the moment, the heat of burning fuel produces explosions that move the parts of an internal combustion engine; or it converts water into steam, the pressure of which turns a turbine to produce electricity. In such devices only 25 to 40 percent of the energy of the burning fuel is turned into useful work; the rest is lost as unused heat. There is little hope that the efficiency can be increased appreciably.

There are, however, other strategies. Burning fuel can heat gases until the atoms and molecules are broken down into electrically charged fragments that can be driven through a magnetic field, thus producing an electric current. Such "magnetohydrodynamic" (MHD) processes would work at substantially higher efficiencies than conventional techniques do.

It is even possible, in theory, to form electricity directly by combining fuel and oxygen in an electric cell without intermediate production of heat. Here efficiencies of 75 percent should be easily attained, with efficiencies of near 100 percent conceivable. So far, practical "fuel cells" have not been devised, yet the difficulties that stand in the way may be defeated.

For that matter, new oil wells may be found. The history of the last half-century is a history of successive predictions of oil exhaustion that did not come true. Before World War II, it was felt that oil production would peak and go into a permanent decline in the 1940s; after the war the date was postponed to the 1960s; now the date has been postponed to 1990. Will it simply continue to be postponed?

Clearly, we can't count on that. What has done most to postpone the day of reckoning has been the discovery of new oil resources from time to time. The largest of these was the rather surprising finding in the years after World War II that the oil reserves of the Middle East were unexpectedly huge. At the present time, 60 percent of the known oil reserves are concentrated into a small area centered about the Persian Gulf (which was also the general site, by a curious coincidence, of humanity's earliest civilization).

It is not likely that we will again come across such another rich find. With each decade, more and more of the Earth has been combed for oil by means of more and more sophisticated techniques. We have found some in northern Alaska, some in the North Sea, we are probing the continental shelves in greater and greater detail—but the day will come and, very likely, soon, when there will be no more oil deposits to find.

With all that we can do by way of conservation, increasing efficiency, and finding new oil wells, it seems inevitable that the twenty-first century will not be very old before all the oil wells will be just about dry. What then?

As it happens, oil can be obtained from sources other than oil wells, in which bits of oil are located in the interstices of underground material, and from which it is relatively easily extracted. There are types of rock called "shale" with which is associated a tarry, organic material called "kerogen." If such shale is heated, the kerogen molecules break down and a substance much like crude oil is obtained. The amount of shale oil in the Earth's crust may be as much as 3,000 times the amount of well oil. One oil-shale field in the western United States may have a total oil content equal to seven times all the oil in the Middle East.

The trouble is that the shale has to be mined; it has to be heated; and the oil produced (and even the richest shale would produce only two barrels per ton of rock) would have to be refined by methods not quite like those now used. Thereafter, the spent shale would have to be disposed of somehow. The difficulties and expenses are very great, and well oil is still too available to force people into underwriting the capital outlay. In the future, though, as well oil declines, shale oil may serve to take up the slack (at, of course, a higher price).

Then, of course, there is coal. Coal was the primary energy source before oil overtook it, and it is still there for the taking. It is commonly said that there is enough coal in the ground to keep the world going at its present rate of energy consumption for thousands of years. However, not all the coal can be obtained by practical mining techniques at the moment. Even so, at the most conservative estimate, coal will last for some hundreds of years and by then mining techniques may have improved.

On the other hand, mining is dangerous. There are explosions, suffocations, cave-ins. The work is physically hard and miners die of lung diseases. The process of mining tends to destroy and pollute the land around the mine and to produce a scene of slag and desolation. Once mined, the coal must be transported; a much more arduous task than that of pumping oil through a pipeline. Coal is far more difficult to handle and ignite than oil is and leaves a heavy ash as well as (unless special efforts are made to clean up the coal before use) an air-polluting smoke.

And yet we may expect that coal will be approached with new and more sophisticated techniques. Land can be restored to something approaching its original condition after mining. (It takes time, labor, and money, to be sure, to do so.) Then, too, much can be done at the mine site, thus avoiding the huge expense and trouble of bulk transportation.

For instance, coal can be burned at the mine site to produce electricity by magnetohydrodynamic techniques. In that case it is the electricity that needs to be transported and not the coal.

Then, too, coal can be heated at the coal mine to give off gases, including carbon monoxide, methane, and hydrogen. These can be so treated as to produce the equivalents of natural gas, gasoline, and other oil products. It will be the oil and gas that will then be transported, rather than the coal, and the coal mines will become our new oil wells.

Even the coal that must be used as coal (in iron and steel manufacture, for instance) can be used more efficiently. It can be reduced to a fine powder, perhaps, that can be transported, ignited, and controlled with little more difficulty than oil can.

Between shale oil and coal mines, we could well have our oil even after the oil wells go dry, and continue our technology essentially as at present for some centuries.

There is, however, a serious difficulty involved in depending on oil and coal, no matter how advanced our techniques. These "fossil fuels" have lain underground for hundreds of millions of years, and they represent many trillion tons of carbon that have all that time not been in Earth's atmosphere in any form.

Now we are burning those fossil fuels at a greater and greater rate, converting the carbon into carbon dioxide and discharging it into the atmosphere. Some of it will dissolve in the ocean. Some of it may be absorbed by the more luxuriant plant growth its presence may encourage. Some of it, however, will stay in the air and raise the atmospheric carbon dioxide content.

Since 1900, for instance, the carbon dioxide content of the atmosphere has risen from 0.029 percent to 0.032 percent. It is estimated that by the year 2000, the concentration may reach a figure of 0.038 percent, an increase of some 30 percent in the century. This must be the result, at least in part, of the burning of fossil fuels, though it may also be due, in part, to the retreat of the forests which are more efficient as carbon dioxide absorbers than are other forms of vegetation.

The increase in carbon dioxide content of the atmosphere is, to be sure, not much. Even if the process of fossil fuel burning continues and accelerates, it is estimated that the highest concentration we are likely to reach would be 0.115 percent. Even this would not interfere with our breathing.

It is not breathing we need worry about, however. It does not take much of an increase in the carbon dioxide concentration to intensify the greenhouse effect appreciably. The average temperature of the Earth could be 1 Celsius degree higher in 2000 than in 1900 because of the added carbon dioxide.* It would take much more than that to reach the point where Earth's climate will be seriously affected and where the ice caps might start melting, with disastrous effects upon the continental lowlands.

There are those, in fact, who point out that if the carbon dioxide content increases above a certain point, the slight rise in the ocean's average temperature will release carbon dioxide from solution in the ocean water, which will further enhance the greenhouse effect, raising the ocean temperature still higher, releasing still more carbon dioxide and so on. Such a "runaway greenhouse effect" might raise Earth's temperature finally to beyond the boiling point of water and make it uninhabitable; and that would surely be a catastrophic consequence of burning fossil fuels.

There is speculation that a short period of mild greenhousing had

---

* To be sure, the greenhouse effect is countered by the fact that industrial activity is also putting more dust into the air. This causes the atmosphere to reflect more sunlight back into space than it ordinarily would, and this would tend to cool the Earth. Indeed, we have had some unusually cold winters in the 1970s. In the end, though, the warming effect of the carbon dioxide is sure to win the race—especially if we take measures to clean up the atmosphere when its pollution reaches dangerous levels.

drastic results in Earth's past. About 75 million years ago, plate tectonics happened to alter Earth's crust in such a way as to cause a number of shallow seas to drain away. Those seas were particularly rich in algae, which served to absorb carbon dioxide from the air. As the shallow seas vanished, the quantity of algae in the ocean declined, and so did the rate of carbon dioxide absorption. The content of atmospheric carbon dioxide therefore increased and the Earth grew warmer.

Large animals are less efficient at losing body heat than are small ones and have greater difficulty in remaining cool. In particular, sperm cells, which are particularly sensitive to heat, may have been damaged at this time so that large animals lost fertility. It may have come about in this way that the dinosaurs became extinct.

Does a similar, and worse, fate—self-inflicted—await us?

In other such cases, I have relied on advances in technology to help us avert or avoid catastrophe, and we might imagine humanity being able to treat the atmosphere in such a way as to extract excess carbon dioxide. However, if the runaway greenhouse effect comes, it is (unlike such catastrophes as a future ice age or an expanding sun) likely to come so soon that it is hard to imagine our technology advancing rapidly enough to save us.

It may well be, then, that projects for finding new oil wells, or for replacing them by shale or coal, are matters of no practical importance; that there is a sharp limit to the rate at which fossil fuel of any kind and from any source can be burned without risking a greenhouse catastrophe. Does that leave us any alternatives or must we wait in despair for civilization to crash one way or another within the next century?

There *are* alternatives. There are the old energy sources that humanity knew before the fossil fuels came on the scene. There are our muscles and those of animals. There is wind, moving water, tides, Earth's internal heat, wood. All of these produce no pollution of consequence, and all of them are renewable and nonexhaustible. What's more, they can be used in more sophisticated fashion than of old.

For instance, we need not chop down trees madly in order to burn them for warmth or to make charcoal for steel manufacture. We can grow special crops cultivated for their speed in absorbing carbon dioxide and building tissue out of them ("biomass"). We might burn those crops directly or, better yet, grow particular varieties from which we can extract inflammable oil or which we can ferment to alcohol. Such naturally produced fuels can help run our future automobiles and factories.

The great advantage of the plant-produced fuels is that they do not involve the permanent addition of carbon dioxide to the air. The fuel is produced from carbon dioxide that has been absorbed some months or years before and that is merely being restored to the atmosphere from which it recently came.

Again, windmills or their equivalent could be built which would work far more efficiently than the medieval structures which inspired them, and which would extract far more energy from the wind.

In older times, tides were used merely to move ships out of harbors. Now they can be used to fill reservoirs at high tide which, at low tide, can produce falling water to turn a turbine and produce electricity. In those areas where Earth's internal heat is near the surface it can be tapped and used to produce the steam that would turn a turbine and generate electricity. There have even been suggestions that we can use the temperature difference between the surface water and the deeper water in the tropical ocean, or the ceaseless energy of the ocean waves, to generate electricity.

These forms of energy are all, by and large, safe and eternal. They do not produce dangerous pollution, and they will always be renewed for as long as Earth and sun are likely to last.

They are not, however, copious. That is, they cannot singly, or even together, supply all of humanity's needs for energy as, over the last two centuries, coal and oil have. That does not mean they are not important. For one thing, each one can, at some given place and time, and for some given purpose, be the most convenient form of energy possible. And all of them together can serve to stretch out the use of the fossil fuels.

With all these other forms of energy available, the burning of fossil fuels can continue at a rate not high enough to endanger the climate, and do so for a long time. During that time, some form of energy which is safe, eternal, and copious can be developed.

The first question is: Does energy with such a combination of characteristics exist? The answer is: Yes, it does.

Energy: Copious

It was only five years after the discovery of radioactivity in 1896 by the French physicist Antoine Henri Becquerel (1852–1908) that

Pierre Curie measured the heat given off by radium as it broke down. That was the first indication that somewhere within the atom there were vast energies of whose existence no one had, till then, had any suspicion.

Almost at once people began to speculate about the possibility of harnessing this energy. The English science-fiction writer H. G. Wells even speculated about the possible existence of what he called "atomic bombs" almost as soon as Curie's discovery had been announced.

It became apparent, however, that in order to release this atomic energy (or, more properly speaking, "nuclear energy," for it was the energy that held the atomic nucleus together, and did not involve the outer electrons that were the basis of chemical reactions) energy had first to be put into the atoms. The atom had to be bombarded with energetic subatomic particles which were positively charged. Few of them hit the nucleus and of those that did, few could overcome the repulsion of the positively charged nucleus and sufficiently disturb its contents to bring about the release of energy. The result was that far more energy had to be put in than could be got out and it seemed that harnessing nuclear energy was a useless dream.

In 1932, however, a new subatomic particle was discovered by James Chadwick (1891–1974). Because it had no electric charge, he called it the "neutron" and, because it had no electric charge, it could approach the electrically charged atomic nucleus without being repelled. It did not require much energy, therefore, to enable a neutron to collide with and enter an atomic nucleus.

The neutron quickly became a favorite subatomic "bullet" and in 1934, the Italian physicist Enrico Fermi (1901–54) bombarded atoms with neutrons in such a way as to change them into atoms of an element one higher up in the list. Uranium was element 92, highest in the list. No element 93 was known and Fermi bombarded uranium, too, in an effort to form the unknown element.

The results were confusing. Other physicists repeated the experiment and tried to make sense of it, notably the German physicist Otto Hahn (1879–1968) and his Austrian co-worker Lise Meitner (1878–1968). It was Meitner who realized late in 1938, that the uranium atom, on being struck by a neutron, split in two ("uranium fission").

She was in exile in Sweden at the time for, as a Jew, she had been forced to leave Nazi Germany. She passed her ideas on to the Danish physicist Niels Bohr (1885–1962) in early 1939 and he brought them to the United States.

The Hungarian-American physicist Leo Szilard (1898–1964) saw the significance of this. The uranium atom, on undergoing fission, released a great deal of energy for a single atom, much more than the small energy of the slow-moving neutron that had struck it. What's more, the uranium atom, as it fissioned, released two or three neutrons, each of which might strike a uranium atom, which would fission, release two or three neutrons each, all of which might strike a uranium atom and so on.

In a tiny fraction of a second, the "chain reaction" that resulted might create an enormous explosion all at the cost of that initial neutron that might be wandering through the air without anyone having taken the trouble to place it there.

He persuaded American scientists to keep quiet about their research (for Germany was about to begin its war against the civilized world) and then persuaded President Roosevelt to support the work by getting Albert Einstein to write a letter on the subject. Before World War II was over, three uranium fission bombs had been constructed. One was tested at Alamogordo, New Mexico, on July 16, 1945, and proved a success. The other two were dropped over Japan.

Meanwhile, though, scientists had devised a way of allowing uranium to undergo fission in a controlled way. The rate of fission reached only a safe level and could be kept going at that level indefinitely. Enough heat would be developed to duplicate the kind of work that burning coal or oil would do, and to produce electricity.

In the 1950s, electricity-producing power stations run by uranium fission were set up in the United States, Great Britain, and the Soviet Union. Since then, such "nuclear fission reactors" have multiplied in many nations, and contribute substantially to satisfying the energy needs of the world.

There are a number of advantages to such nuclear fission reactors. For one thing, weight for weight, uranium produces far more energy than burning coal or oil do. In fact, even though uranium is not exactly a common metal, the world's supply, it is estimated, will produce ten to a hundred times as much energy as will its stores of fossil fuel.

One of the reasons that uranium doesn't do better still is that there are two varieties of uranium, only one of which will undergo fission. The varieties are uranium-238 and uranium-235 and it is only uranium-235 that undergoes fission under the conditions of bombardment by slow neutrons. As it happens, uranium-235 makes up only 0.7 percent of the uranium found in nature.

It is possible, however, to design a nuclear fission reactor in such a way that the fissioning core is surrounded by ordinary uranium-238, or by the similar metal, thorium-232. Neutrons leaking out of the core will strike the uranium or thorium atoms and, while not causing them to undergo fission, will change them into other types of atoms which, under the proper conditions, *will* do so. Such a reactor breeds fuel in the form of fissionable plutonium-239 or uranium-233, even as its original uranium-235 fuel is slowly consumed. In fact, it breeds more fuel than it consumes and it is called a "breeder reactor" in consequence.

So far, almost all the nuclear fission reactors in use are not breeders, but some breeder reactors have been built, even as early as 1951, and more can be built at any time. With the use of breeder reactors, all the uranium and thorium in the world can be fissioned and made to produce energy. In this way, humanity will have available to it a source of energy at least 3,000 times as great as the entire fossil fuel store.

Using ordinary nuclear fission reactors, humanity will have a store of energy that will last for centuries at present rates of use. With breeder reactors, the store of energy will last for hundreds of thousands of years—plenty of time to work out a still better strategy long before it runs out. What's more, nuclear fission reactors, whether ordinary or breeder, do not produce carbon dioxide or any form of chemical air pollution.

With these advantages, what disadvantages can there be? To begin with, uranium and thorium are well scattered through the Earth's crust and are hard to find and concentrate. It may be that only a rather small fraction of the uranium and thorium that exist can be used. Second, the nuclear fission reactors are large expensive devices that are not easy to maintain and are hard to repair. Third, and most important, the nuclear fission reactors introduce a new and particularly deadly form of pollution—that of hard radiation.

When the uranium atom fissions it produces a whole series of smaller atoms that are radioactive, far more intensely radioactive than uranium itself. This radioactivity falls off to a safe level only slowly, in the case of some varieties only after thousands of years. This "radioactive ash" is highly dangerous since its radiations can kill, as surely as a nuclear bomb will, though far more insidiously. If humanity's energy needs were filled by fission reactors exclusively, the amount of radioactivity present in the ash produced each year would be the equivalent of millions of fission-bomb explosions.

The radioactive ash has to be stored in some safe place in such a way that it will not get into the environment for thousands of years. It can be stored in stainless steel containers, or it can be mixed with melted glass that is then allowed to congeal. The containers, or glass, can be stored in underground salt mines, in Antarctica, in sediments of the ocean bottom and so on. So far, numerous methods of disposal have been proposed, all of them with some credibility, but none of them sufficiently safe to satisfy everybody.

Then, too, it is always possible that a nuclear fission reactor may go out of control. The reactor is so designed that it is impossible for it to blow up, but sizable quantities of fissioning material must be used and if the fission reaction accidentally speeds up past the safety melting point, the core will heat up, melt through its protective coatings, and deadly radiation may blow over a large area.

Breeder reactors are felt by some to be particularly deadly because the fuel they use is frequently the metal plutonium, which is more radioactive than uranium, and which retains that radioactivity for hundreds of thousands of years. It is considered by some to be the deadliest substance on Earth, and there are fears that if plutonium becomes too common, it may escape into the environment and literally poison the Earth into uninhabitability.

There is also the fear that plutonium may exalt terrorism to new heights of effectiveness. If terrorists could obtain a supply of plutonium, they could use the threat of explosion or poison with which to blackmail the world. It would be a far more terrible weapon than anything available to them up to this point.

There is no way of assuring people that such things will *never* happen and, as a result, there are more and more objections to the building of nuclear fission reactors. Nuclear fission power is expanding far more slowly than had been anticipated in the 1950s when it came into use amid glowing predictions of a new age of energy-plenty.

And yet fission is not the only route to nuclear energy. In the universe generally, the main source of energy is produced by the fusion of hydrogen nuclei (the simplest that exist) into helium nuclei (the next simplest). It is this "hydrogen fusion" which powers the stars, as was pointed out by the German-American physicist Hans Albrecht Bethe (1906–) in 1938.

After World War II, physicists attempted to produce hydrogen fusion in the laboratory. For that they needed extreme temperatures in the millions of degrees, and they had to hold the hydrogen gas in place while it was being raised to such an enormously high temperature. The sun and other stars held their cores in place by enormous gravitational fields, but on Earth that could not be duplicated.

One way out was to raise the temperature of hydrogen so rapidly that it would not have time to expand and drift away before it was hot enough to fuse. A nuclear fission bomb would do the trick, and in 1952, a bomb was exploded in the United States in which fissioning uranium set off fusing hydrogen. The Soviet Union immediately followed with one of its own.

Such a "nuclear fusion bomb" or "hydrogen bomb" was enormously more powerful than the fission bombs and they have never been used in war. Because fusion bombs require high temperatures for their working, they are also called "thermonuclear bombs" and it is their use in a "thermonuclear war" that I have considered as possibly bringing about a catastrophe of the fourth class.

But could hydrogen fusion come under control and produce energy as tamely as uranium fission does? The English physicist John David Lawson (1923–) worked out the requirements in 1957. The hydrogen would have to be of a certain density, reach a certain temperature, and hold that temperature without escaping for a certain length of time. Any fall-short in one of these properties requires an increase in one or both of the others. Ever since, scientists in the United States, Great Britain, and the Soviet Union have been attempting to meet those conditions.

There are three types of hydrogen atoms, hydrogen-1, hydrogen-2, and hydrogen-3. Hydrogen-2 is called "deuterium" and hydrogen-3 is called "tritium." Hydrogen-2 fuses at a lower temperature than hydrogen-1, and hydrogen-3 fuses at a lower temperature still (though even the lowest temperature for fusion is still in the tens of millions of degrees, under earthly conditions).

Hydrogen-3 is a radioactive atom that hardly exists in nature. It can be made in the laboratory, but it can only be used on a small scale. Hydrogen-2 is therefore the prime fusion fuel, and a bit of hydrogen-3 is added to lower the fusion temperature.

Hydrogen-2 is much less common than hydrogen-1. Out of every 100,000 atoms of hydrogen only 15 are hydrogen-2. Even so, there is enough hydrogen-2 present in one gallon of sea water to represent the energy obtained by burning 350 gallons of gasoline. And the ocean (within which two atoms out of every three are hydrogen) is so vast that it contains enough hydrogen-2 to keep producing energy at the present rate of world use for billions of years.

There are a number of respects in which nuclear fusion seems preferable to nuclear fission. For one thing, weight for weight there is about ten times as much energy to be gotten out of matter by fusion than by fission, and hydrogen-2, the fuel of fusion, is much easier to get than either uranium or thorium, and much easier to handle. Once hydrogen-2 is set to fusing, only microscopic quantities of it will be used at any one time, so that even if the fusion goes out of control and the entire fusable material goes at once, the results would be only a very minor explosion, not enough to notice. Furthermore, hydrogen fusion does not produce any radioactive ash. Its main product, helium, is the least dangerous substance known. In the course of fusion, hydrogen-3 and neutrons are produced and both are dangerous. They are, however, produced in minor quantities and they can be recycled and used in the course of further fusion.

In every respect then, nuclear fusion would seem to be the ideal energy source. The catch is, though, that we don't have it yet. Despite years of trying, scientists have not yet held enough hydrogen in place at a high enough temperature for a long enough time to allow it to fuse under controlled conditions.

Scientists are approaching the problem from several directions. Strong, carefully designed magnetic fields hold the charged fragments in place, while the temperature is raised slowly. Or else the temperature is raised very rapidly, not with fission bombs, but with laser light or electron beams. There seems to be a reasonable chance that during the 1980s one of these methods will work, or possibly all three, and that controlled fusion in the laboratory will be a fact. It may then take some decades to build large fusion power stations that can contribute substantially to humanity's energy needs.

Leaving hydrogen fusion to one side, however, there is one other source of copious energy that is safe and eternal, and that is solar radiation. Two percent of the energy in sunlight supports the photosynthesis of all plant life on Earth and through that all animal life. The rest of the energy in sunlight is at least ten thousand times as great as all the energy needs of humanity. This major portion of the sun's radiation is far from useless. It vaporizes the ocean and therefore produces rain, running water, and Earth's fresh-water supply in general. It supports ocean currents and the wind. It warms the Earth generally and makes it habitable.

Nevertheless, there is no reason why human beings can't make use of the sun's radiation first. When we do, the net result is that the radiation is turned to heat and nothing is lost. It would be like stepping under a waterfall: the water would still reach ground level and move downstream but we would have interrupted enough of it, temporarily, to wash and refresh ourselves.

To be sure there is a major difficulty in that solar energy, while copious, is dilute. It is spread thinly over a wide area and collecting it and making use of it would not be easy.

On a small scale, solar energy has been used for a long time. Southern windows in winter let in sunlight and are relatively opaque to the reradiation of infrared light, so that a house is warmed by the greenhouse effect and needs less fuel.

More can be be done after that fashion. Watertanks on southern-sloping roofs (northern-sloping in the southern hemisphere) can absorb heat from the sun and supply a house with a perpetual hot-water source. This could also be used to warm the house generally, or, for that matter, to air condition it in summer. Or solar radiation can be converted directly to electricity by exposing solar cells to sunlight.

To be sure, sunlight is not perpetually available. There is none at night and, even during the day, clouds can reduce the light to a useless level. There is also the point that at various times of the day, a house can be shaded by other houses or by natural objects such as hills and trees. Nor is there any completely adequate means of storing solar energy during periods of brightness for use in times of darkness.

If solar energy is to run the world, rather than dealing with individual houses here and there, it could be necessary to coat tens of thousands of square miles of desert area with solar cells. This would be very expensive to install and maintain.

There is the possibility, however, of collecting solar energy, not on Earth's surface, but in nearby space. A wide bank of solar cells placed in orbit in the equatorial plane about 33,000 kilometers (21,000 miles) above the Earth's surface, will revolve about the Earth in twenty-four hours. This is a "synchronous orbit" and the space station will seem to be motionless with respect to Earth's surface.

Such a bank of solar cells would be exposed to the full range of the sun's radiation without any atmospheric interference. It will be in the Earth's shadow only about 2 percent of the time in the course of a year, thus greatly reducing the necessity of energy storage. Some estimates have it that a given area of solar cells in synchronous orbit will produce sixty times as much electricity as that same area on Earth's surface.

The electricity formed in the space station would be converted to microwave radiation beamed down to a receiving station on Earth and there reconverted to electricity. A hundred such stations, dotted round the equatorial plane, would represent a source of copious energy that would last as long as the sun does.

If we look into the future on the assumption that human beings will cooperate for survival, it may seem that by 2020 not only will there be nuclear fusion power stations in operation, but the first few solar power space stations will also be in operation. We can certainly make it to 2020 by using fossil fuels and other energy sources. Given peace and goodwill, then, the energy crisis that now afflicts us may prove to be no crisis at all in the long run. Furthermore, the exploitation of space in connection with solar energy stations would lead to far more than that. Laboratories and observatories will also be built in space, together with space settlements to house the people who will be doing the building. There will be mining stations on the moon to supply most of the material resources for the space structures (though carbon, nitrogen, and hydrogen will have to continue to be supplied by Earth for a while).

Eventually, much of Earth's industrial plant will be moved into space; the asteroids will be mined; and humanity will begin to expand throughout the solar system and, in time, even toward the stars. With such a scenario, we might suppose that all problems will be solved —except that the very victory will itself bring problems. It is to the possible catastrophes arising from victory that I will address myself in the final chapter.

Chapter 15

The Dangers of Victory

Population

If we imagine a society at peace, with plentiful energy and, therefore, with abundant capacity to recycle resources and to advance technology, we must also imagine that that society will reap the rewards of its victory over the environment. The most obvious reward will be precisely that which has been experienced as a result of similar victories in the past—the increase of population.

The human species, like all living species that have existed on Earth, has the capacity for a rapid increase in its numbers. It is not impossible for a woman to have, let us say, sixteen children during the years in which she is capable of childbearing. (Cases of over thirty children by a single mother have been reported.) That means that if we begin with two people, a man and wife, we will have a total of eighteen people after thirty years. The older children might by then have intermarried among themselves (if we imagine a society that permits incest) and have produced some ten more children. From two to twenty-eight, then—a fourteenfold increase in thirty years. At that rate, the original pair of human beings would have increased to 100 million in two centuries.

The human population does not increase at any such rate, however, and never has, for two reasons. In the first place, the number of births is not a universal sixteen to a woman, but averages considerably fewer for a variety of reasons. In other words, the birthrate generally falls below its potential maximum.

In the second place, I have been assuming that all people who are born remain alive and, of course, that is not so. All people must eventually die; very often before they have brought about the birth of as many young as they might; sometimes before they have brought about the birth of any young at all.

In short, there is a deathrate as well as a birthrate and for most species at most times, the two are fairly equal.

If, in the long run, the deathrate and the birthrate remain equal, then the population of any species under discussion remains stable, but if the deathrate rises higher than the birthrate, even if only slightly, then the species dwindles in numbers and eventually becomes extinct. If the birthrate remains even only slightly higher than the deathrate then the species will increase steadily in number.

The deathrate for any species tends to climb if the environment turns unfavorable to them for any reason, and to drop if it turns favorable. The population of any species tends to rise in good years and sink in bad years.

Human beings alone, of all the species that have lived on Earth, have had the intelligence and the opportunity to alter their environment radically in such a way as to favor themselves. They have improved their climate by the use of fire, for instance; increased their food supply by the deliberate cultivation of plants and herding of animals; have, through the invention of weapons, reduced the danger from predators; and through the development of medicine, reduced the danger from parasites. The result has been that humanity has been able to maintain a birthrate that has, on the whole, been higher than the deathrate ever since *Homo sapiens* first appeared on the planet.

By 6000 B.C., when agriculture and herding were yet in their infancy, the total human population on Earth had risen to 10 million. At the time of the building of the Great Pyramid, it was probably something like 40 million; in the time of Homer, 100 million; in the time of Columbus, 500 million; in the time of Napoleon, 1 billion; in the time of Lenin, 2 billion. And now, in the 1970s, the human population has reached the 4 billion mark.

Since technology tends to be cumulative, the rate at which humanity has been increasing its domination over the environment and over competing life forms, and the rate at which physical security has been advancing, has been increasing steadily. This means that the disparity between birthrate and deathrate has been increasingly favoring the former. This, in turn, means that not only is the human population increasing, but it has been doing so at a steadily advancing rate.

In the millennia before agriculture, when human beings lived by hunting and food gathering, the food supply was skimpy and insecure and humanity could increase its numbers only by spreading out more widely over the face of the Earth. The rate of population increase, then, must have been less than 0.02 percent per year, and it must have taken more than 35,000 years for the human population to double in size.

With the development of agriculture and herding, and the assurance of a more stable and copious food supply, and with other technological advances, the rate of population increase began to rise, reaching 0.3 percent per year in 1700 (a doubling period of 230 years) and 0.5 percent per year in 1800 (a doubling period of 140 years).

The coming of the Industrial Revolution, of the mechanization of agriculture, and the rapid advance of medicine, further lifted the rate of population increase to 1 percent per year in 1900 (a doubling period of 70 years) and 2 percent per year in the 1970s (a doubling period of 35 years).

The increase in both population *and* the rate of population increase, multiplies the rate at which new mouths are added to humanity. Thus, in 1800, when the total population was 1 billion and the rate of increase was 0.5 percent per year, that meant 5 million new mouths to feed each year. In the 1970s, with the population 4 billion and the rate of increase 2 percent per year, there are 80 million new mouths to feed each year. The population in 170 years has increased fourfold, but the additional numbers each year sixteenfold.

Although this is all a testimony to human triumph over the environment, it is also a terrible threat. A declining population can decline indefinitely until it reaches the ultimate figure of zero. An increasing population, however, cannot, *under any circumstances*, continue to increase indefinitely. Eventually, an increasing population will outpace its food supply, outpace its environmental requirements, outpace its living space, and then, with what would very likely be catastrophic speed, the situation would be reversed and there would be a sharp population decline.

Such a population boom-and-bust has been observed in numerous other species, who have multiplied exceedingly in a succession of years in which the climate and other aspects of the environment have, by chance, favored their growth—only to die in hordes as the inevitable bad year cut their food supply.

This is the population doom that faces humanity, too. The very victory that increases our population will bring us to a height from which we will have no choice but to tumble—and the greater the height, the more disastrous the tumble.

Can we count on technological advances warding off the evil in the future as it has done in the past? No, for it is easy to show with absolute certainty that the present rate of population growth, *if it continues*, will easily outpace not only any likely technological advance, but any conceivable technological advance.

Let us begin with the fact that the Earth's population is 4 billion in 1979 (actually a bit higher) and that the rate of population is, and will continue to be, 2 percent per year. We might argue that a human population of 4 billion is already too high for the Earth to endure and never mind any increase. Some 500 million people, an eighth of the whole (mostly in Asia and Africa), are chronically and seriously undernourished, and hundreds of thousands die of starvation each year. Furthermore, the pressures of producing more food each year to feed more mouths has forced human beings to place marginal lands under cultivation, to use pesticides, fertilizers, and irrigation not wisely but too much, and to upset the ecological balance of Earth more and more drastically. In consequence, soil is being eroded, deserts are advancing, and food production (which has been rising with the population, and even somewhat faster, in these last desperate decades of population explosion) is approaching a plateau and may soon begin to decline. In that case, famine will become more widespread each year.

On the other hand, it might be argued that food shortages are man-made, the result of waste, inefficiency, greed, and injustice. With more humane and better governments, more sensible land use, more thrifty life patterns, more equitable food distribution, Earth could support a far larger population than today's without unduly burdening its capacity. The largest figure that has been offered is 50 billion, or 12½ times the present population.

At the present 2 percent per year rate of increase, however, the Earth's population will double every 35 years. In 2014, it will be 8 billion; in 2049, it will be 16 billion, and so on. This means that at the present rate of increase the Earth's population will be 50 billion in about 2100, only 120 years from now. And then what? If, having reached that point, we *then* outpace the food supply, the sudden bust will be that much more catastrophic.

Of course, in 120 years human technology will have worked out new ways of feeding humanity—by wiping out all other forms of animal life, and cultivating plants that are one hundred percent edible, and then living on those plants without competition. In that way, Earth might be made to support 1.2 trillion, or 300 times the present population. And yet, at the present rate of increase, a population of 1.2 trillion will be reached in 2280, just about 300 years from now. Then what?

It is, in fact, pointless to argue that there are specific numbers of people whom we can support through this scientific advance or that. A geometric progression (which is what population increase is) can outpace any number. Let us reason it out.

Suppose that the average weight of a human being (women and children included) is 45 kilograms (100 pounds). In that case, the total mass of humanity now living on Earth would weigh 180 billion kilograms. This weight would double every 35 years as the population doubles. At this rate of increase, if one carries matters to the extreme, in 1,800 years, the total mass of humanity would equal the total mass of the Earth. (This is not a long time-lapse. It is only 1,800 years since the time of the Emperor Marcus Aurelius.)

No one can possibly suppose that humanity on Earth can multiply its numbers until the ball of the planet is solid human flesh and blood. This means, in fact, that *no matter what we do,* we cannot maintain our present population increase on Earth for more than 1,800 years.

But why restrict ourselves to Earth? Long before 1,800 years has elapsed, humanity will have reached other worlds and constructed artificial space settlements, and both could accommodate our increasing numbers. One might even maintain that by taking over the universe, the total mass of human flesh and blood might indeed exceed the mass of the Earth some day. Yet even that cannot withstand the power of a geometric progression.

The sun is 330,000 times as massive as Earth, and the Galaxy is 150 billion times as massive as the sun. There may be as many as 100 billion galaxies in the universe altogether. If we suppose that the average galaxy is as massive as our own (an overestimate almost certainly, but never mind) then the total mass of the universe is 5,000,000,000,000,000,000,000,000,000 times that of the Earth. And yet, if the present human population continues to increase at a steady 2 percent per year, the total mass of human flesh and blood will equal the mass of the universe in a little over 5,000 years. This is about the time that has elapsed since writing was invented.

In other words, during the first 5,000 years of written history, we have reached the stage where we somewhat crowd the surface of one small planet. During the next 5,000 years, at the present rate of increase, we will run out of not merely that planet, but of the entire universe.

It follows, then, that if we are to avoid outracing our food supplies, our resources, our room, we must halt the present rate of population increase in less than 5,000 years even if we imagine ourselves advancing our technology to the uttermost limits of fantasy. And if we are to be honestly realistic about it, we can only have a fair chance of avoiding a catastrophe of the fifth class if we start reducing the rate of population growth *now!*

But how? It is indeed a problem, for in all the history of life, no species has attempted to control its own numbers voluntarily.* Not even the human species has attempted to do so. It has, until now given birth to offspring freely, and increased its numbers to the limit of the possible.

In order to control population now, the disparity between birthrate and deathrate must somehow be reduced, and the growing predominance of the former over the latter must be decreased. To achieve a stationary population, or even a temporarily declining one, we have only two alternatives: either the deathrate must be increased until it matches or exceeds the birthrate, or the birthrate must be decreased until it matches or falls short of the deathrate.†

Raising the deathrate is the easier alternative. Among all species of plants and animals, throughout the history of life, a sudden and dramatic increase in deathrate has been the usual answer to a population

---

* Experiments with rats have shown that extreme overcrowding induces such a psychotic society that young are not produced or, if produced, are not cared for. This, however, is not voluntary control, and for human beings to wait for crowding to become extreme enough to madden society, is to wait for catastrophe.

† It is possible to combine the two, and both to raise the deathrate and decrease the birthrate.

increase that has carried the species upward to a level insupportable in the long run. The deathrate is increased primarily as a result of starvation. The weakening that precedes starvation makes it easier for individuals of the species to fall prey to disease and predation as well.

In the case of human beings in past history, the same can be said, and if we look into the future we can count on our population being controlled (if all else fails) by starvation, disease, and violence—all followed by death. That this is not a new idea can be attested to by the fact that these four—starvation, disease, violence, and death—are the Four Horsemen of the Apocalypse, described in the biblical Book of Revelation as bedeviling humanity in its last days.

It is clear, however, that to solve the population problem by increasing the deathrate is simply to experience a catastrophe of the fifth class in which civilization breaks down. If, in the squabbling over the last dregs of food and resources, a thermonuclear war is elicited as a measure of desperation, a catastrophe of the fourth class may follow and humanity may be wiped out.

We are left, then, with a lowering of the birthrate as the only way of avoiding catastrophe. But how?

Controlling the birthrate after the fact, as by infanticide, or even by abortion, is repugnant to many people. Even if it is not made a question of the "sanctity of life" (a principle never given more than lip-service in human history) we might ask why a woman should have to undergo the discomfort of a pregnancy only to have the result destroyed, or why she should have to undergo the discomfort of an abortion? Why not simply prevent conception in the first place!

One foolproof way of avoiding conception is by avoiding sex, but there is every reason to think that this will never be a popular method of controlling the population. Instead, it is necessary to uncouple sex and conception, making it possible to have the former without the latter, except where children are actually desired, and where they are necessary to maintain a tolerable population level.

For contraception, there are a variety of methods, surgical, mechanical, and chemical, all of which are well known and which need only to be applied intelligently. There are, in fact, well-known varieties of sexual activity that are fully enjoyed and practiced that do no detectable harm to the practitioners or to anyone else, and that carry with them absolutely no chance of conception.

There is thus no practical difficulty in lowering the birthrate—only social and psychological ones. Society has been accustomed for so long to a surplus of children (because of the high deathrate among them) that in many places the economy, and in almost all places the individual psychology, depends on it. Contraception is bitterly fought by many traditionalist groups as being immoral, and many children per family is something that is still viewed, traditionally, as a blessing.

What will happen, then? With salvation possible, will humanity slide down the chute to catastrophe simply out of habituation to an outmoded way of thought? It is possible that this is exactly what might happen. And yet, more and more people (such as I, myself) have been speaking and writing about the population danger, and about the visible destruction of the environment brought on by the increasing load of humanity and by the increasing demands of more and more people for more food, more energy, and more amenities of life. Increasingly, then, government leaders are beginning to recognize that *no* problem can be solved as long as the population problem is not solved, and that *all* causes are lost causes while the population continues to grow. As a result there is an increasing drive, in one way or another, for a decrease in the birthrate. This is an extraordinarily hopeful sign, for social pressure can do more to reduce the birthrate than anything else.

Apparently, as the 1970s waned, the birthrate in the world *was* declining, and the rate of population increase dropped from 2 percent to 1.8 percent. This is not enough, of course, for at the present moment *any* increase will bring with it an eventual catastrophe if that increase continues. Nevertheless, the decrease is a hopeful sign.

It may be, then, that though the population will continue to increase it will do so at a decreasing rate, reach a maximum of perhaps no more than 8 billion, and will then decline. The process will create damage enough, but it is possible that civilization will weather the storm and that humanity, much battered, may survive, repair the Earth and its ecological balance, and rebuild a wiser and more practical culture based on a stable population held to a tolerable number.

Education

We may envisage a time, then, say a hundred years from now, in which the population problem is solved, in which energy is cheap and abundant, in which humanity recycles its resources and maintains itself in peace and serenity. Surely, then, all problems will be solved, and all catastrophes avoided.

Not necessarily. Every solution is bound to bring in the wake of its victory its own problems. A world in which population is controlled is one in which the birthrate is as low as the deathrate, and since thanks to modern medicine the deathrate is far lower now than it has ever been in the past, so must the birthrate be. This means that, on a percentage basis, there will be fewer babies and young people than there have ever been, and a greater number of mature and elderly people. Indeed, if we imagine medical technology advancing, the average lifespan could continue to increase. That would mean the deathrate would continue to fall—and the birthrate would have to fall with it.

The kind of society we must anticipate, then, if we are to achieve a stable population, is one with an advancing median age. We will witness, so to speak, the "graying of Earth." We can actually see this happening in those portions of the world where the birthrate has dropped and life expectancy has lengthened—in the United States, for instance.

In 1900, when the life expectancy in the United States was only about 40 years, there were 3,100,000 people over 65 out of a total population of 77 million, or just about 4 percent. By 1940, there were 9 million people over 65 out of a total population of 134 million, or 6.7 percent. In 1970, there were 20. 2 million people over 65 out of a total population of 208 million, or nearly 10 percent. By the year 2000, there may be as many as 29 million people over 65 out of an estimated 240 million, or 12 percent. In a hundred years, while the population will have a little more than tripled, the number over 65 will have increased nearly tenfold.

The effect on American politics and the economy is clear. The elderly are an increasingly potent part of the electorate and the nation's political and financial institutions must concern themselves more and more with pensions, social security, medical insurance, and so on.

To be sure, everyone wants a long life and everyone wants to be cared for when old, yet from the standpoint of civilization as a whole, there might be a problem. If, as a result of population stabilization, we develop an aging humanity, might it not be that the spirit, adventure, and imagination of youth would dwindle and die under the stodgy conservatism and fullness of age? Might not the burden of innovation and daring rest on so few that the dead weight of the old would break down civilization? Might not civilization, having evaded the death-by-bang of a population explosion, find itself suffering the death-by-whimper of population aging.

But are age and stodginess necessarily linked? Ours is the first society that takes this for granted, for ours is the first society in which the aged have grown superfluous. In semiliterate nonrecord-keeping societies, the old were the repositories and guardians of traditions, the living reference books, libraries, and oracles. Nowadays, however, we don't need the memories of the old; we have far better ways of keeping records. As a result, the aged lose their function and their grip on our respect.

Again, in societies in which technology changed slowly, it was the old artisan, rich in experience, in knowledge, who could be depended on for the skilled eye, the shrewd judgment, and the good job. Now technology changes rapidly and it is the downy-cheeked college graduate we want, expecting him to bring with him the latest techniques. To make room for him, we forcibly retire the old, and again, age loses its function. And as the numbers of functionless aged increases, they seem indeed to be a deadweight. But must they be?

People today live on the average, twice as long as did our ancestors of a century and a half ago. Long life, however, is not the only change. People today are also healthier and stronger, on the average, at any given age than were their ancestors at that same age.

It was not just that people died young in the days before modern medicine. Many of them were also visibly old at thirty. Living that long or longer meant having to survive repeated bouts of infectious disease that we can now either prevent or easily cure. It meant living on deficient diets, both quantitatively and qualitatively. There was no way of fighting diseased teeth or chronic infections, no way of ameliorating the effects of hormone malfunctioning or vitamin deficiency, no way of countering dozens of other disabilities. To top it off, many people had to exhaust themselves with the kind of unending toil that today machines do for us.*

As a result, the aging persons of today are vigorous and young compared to those of identical years in the medieval days of chivalry and even in the United States of the pioneers.

---

* Many people today dream of a past in which people "lived close to nature" and were healthy and hearty beyond today's city-crowded, pollution-riddled people. Such dreamers would be unpleasantly surprised were they to find themselves in the *real* past—disease-ridden, famished, and filthy, even at the highest levels.

We can assume that this trend toward more vigorous older people will continue into the future if civilization survives and medical technology advances. The whole concept of "youth" and "age" may become rather blurred in the stable-population-to-come. But then, even if the physical difference between youth and age diminishes, what about the mental differences? Can anything be done about the stagnation of age, its inability to accept creative change?

But how much of this stagnation of age is created by the traditions of a youth-centered society? Despite the gradual extension of the period of schooling, education continues to be associated with youth, and continues to have a kind of cutoff date. There continues to be a strong feeling that there comes a time when an education is *completed,* and that this time is not very far along in a person's lifetime.

In a sense, this lends an aura of disgrace to education. Most young people who chafe under the discipline of enforced schooling and the discomforts of incompetent teaching, can't help but notice that grown-ups need not go to school. One of the rewards of adulthood, it must surely seem to a rebellious youngster, is that of casting off the educational shackles. To them the ideal of outgrowing childhood is to reach a state of never having to learn anything again.

The nature of education today makes it inevitable that it be viewed as the penalty of youth, and that puts a premium on failure. The youngster who drops out of school prematurely and who abandons further education to take some sort of immediate job appears to his peers to have graduated to adulthood. The adult, on the other hand, who attempts to learn something new is often looked on with a vague amusement by many, and is considered as somehow betraying a second childishness.

By equating education only with youth and by making it socially difficult for the average person to learn after the days of formal schooling are over, we make sure that most people are left with nothing more than the information and attitudes gained in teen-age years, and vaguely remembered—and then we complain of the stodginess of age.

This shortcoming of education with respect to the individual may be overshadowed by another shortcoming with respect to society as a whole. It may be that all of society may be forced to stop learning. Can it be that the progress of human knowledge will be forced to halt simply through its own superlative success? We have learned so much that it is becoming difficult to find the specific items we need among the vast mass of the whole, specific items that may be crucial to further advance. And if humanity can no longer stumble forward on the road of scientific and technological advance, will we no longer be able to maintain our civilization? Is this another of the dangers of victory?

We might summarize the danger by saying that the sum total of human knowledge lacks an index, and that there is no efficient method of retrieval of information. How can we correct this but by calling on a more-than-human memory to serve as an index, and a faster-than-human system of retrieval to make use of the index?

In short, we need a computer, and for nearly forty years, we have been developing better, faster, more compact and more versatile computers at a breakneck pace. This trend should continue if civilization remains intact, and in that case the computerization of knowledge is inevitable. More and more information will be recorded on microfilm and more and more of that will be accessible by computer.

There will be a tendency to centralize information so that a request for particular items can tap the resources of all the libraries of a region, or of a nation, and, eventually, of the world. There will be the equivalent, at last, of a Global Computerized Library in which the total available knowledge of humanity will be stored and from which any item of that total can be retrieved on demand.

The manner in which such a library would be tapped is no mystery; the technique is on the way. We already have communications satellites that make it possible to connect any two points on the globe in a matter of fractions of a second.

Present-day communications satellites depend on radio waves for interconnection, however, and the number of possible channels that they make available is sharply limited. In future generations of such satellites, lasers, making use of visible light and of ultraviolet radiation, will be used for the interconnection. (The first laser was constructed as recently as 1960 by the American physicist Theodore Harold Maiman (1927–).) The wavelengths of visible light and ultraviolet radiation are millions of times shorter than those of radio waves, so that the laser beams could carry millions of times as many channels as radio beams can.

The time could come, then, when every human being would be assigned a specific television channel of his own, which could be tuned to a computer outlet that would be his or her connection with the gathered knowledge of the world. The equivalent of a television set would produce wanted material on a screen, or would reproduce it on film or paper—stock-market quotations, news of the day, shopping opportunities, parts or all of a newspaper, magazine, or book.

The Global Computerized Library would be essential for scholars and for research, but this would represent a minor fraction of its use. It would represent an enormous revolution in education and, for the first time, offer us a scheme of education that would be truly open to all people of any age.

People after all *want* to learn. They have a three-pound brain in each skull which demands constant occupation to prevent the painful disease of boredom. In default of anything better or more rewarding, it can be filled with the aimless visions of low-quality television programs or the aimless sounds of low-quality recordings.

Even this poor material is preferable to schools as presently constituted, where the individual students are mass-fed certain stereotyped subjects at certain dictated speeds, without any regard for what it is the individual wishes to know and for how rapidly or slowly he can absorb the information.

What if, however, there were a device in a person's living quarters that would feed information to him or her on exactly what he or she wants to know: how to build a stamp collection, how to mend fences, how to bake bread, how to make love, details on the private lives of the kings of England, the rules of football, the history of the stage? What if all this were presented with endless patience, with endless repetition if necessary, and at a time and place of the learner's own choosing?

What if, having absorbed some of a subject, the learner were to ask for something more advanced, or a little to the side? What if some item in the information happened to fire a sudden new interest and sent the learner off in a completely new direction?

Why not? Surely more and more people would take this easy and natural way of satisfying curiosity and the desire to know. And each person, as he is educated in his own interests, could then begin to make contributions of his or her own. The person who had a new thought or observation of any kind in any field could report it, and if it did not duplicate something already in the library, it could be held for confirmation and, possibly, be added, eventually, to the common store. Each person would be a teacher as well as a learner.

With the ultimate library the ultimate teaching machine as well, would the teacher-learner lose all desire for human interaction?

Would civilization develop into a vast community of isolates, and would it break down in that fashion? Why should it? No teaching machine could replace human contact in all areas. In athletics, in public speaking, in the dramatic arts, in exploration, in dancing, in lovemaking—no amount of bookishness would replace practice, though theory might improve it. People would still interact, and all the more intricately and pleasurably for knowing what they are doing.

In fact, we can rely on every human being to possess a missionary instinct in connection with whatever matter he or she is devouringly interested in. The chess enthusiast tries to get others interested in chess and the same can be said, analogously, of fishermen, dancers, chemists, historians, joggers, antique-buyers, or anyone else. The person who probes the teaching machine and finds a fascination in weaving, or in the history of costumes, or in Roman coins, would very likely make a determined effort to find others of like interests.

And this method of education by computer would surely be no respecter of age. It could be used by anyone at any age, with new interests starting in the sixties, perhaps, and old interests fading. Constant exercise of curiosity and thought would keep the brain as supple as constant exercise would keep the body in shape. It would follow, then, that stodginess need not accompany advancing years; at least not as soon and not as surely.

The result could well be that despite the unprecedented aging of the human population and the never-before-seen underrepresentation of youth, the world of stable population would be one of rapid technological advance and an unparalleled intensity of intellectual cross-fertilization.

But might not even the new free-will education bring dangers in its wake? With everyone free to learn as he or she wishes, would not almost everyone follow the tracks of trivia? Who would learn the dull, hard things that would be required to run the world?

In the computerized world of the future, however, it is precisely the really dull things that would no longer be the province of human beings. Automatic machinery would take care of it. To the human beings would be left those creative aspects of the mind that would come under the heading of amusement to those involved with them.

There will always be those who would find amusement in mathematics and science, in politics and business, in research and development. They would help "run" the world, but would be doing so as much out of desire and pleasure as were those who were occupied in the building of rock gardens or the devising of gourmet recipes.

Would those running the world enrich themselves and oppress others? Presumably that possibility remains, but one can hope that in a properly computerized world, the chance of corruption would at least be smaller and that a smoothly running world would bring more benefits to people generally than corruption plus disorder could bring to a few.

The picture of a utopia arises. It would be a world in which national rivalries would be detoxified and war abolished. It would be a world in which racism, sexism, and ageism would lose their importance in a cooperative society of advanced communication, automation, and computerization. It would be a world of copious energy and flourishing technology.

But can even utopia have its dangers? After all, in a world of leisure and amusement, might not the inner fiber of humanity relax, soften, and decay. *Homo sapiens* has developed and grown strong in an atmosphere of continual risk and danger. Once Earth is converted into a global Sunday afternoon in the suburbs, might civilization, having avoided the death-by-bang of population explosion, and the death-by-whimper of population aging, fall prey to a death-in-silence by boredom?

Perhaps so, if Earth were all there was, but it would seem certain that by the time such suburbanity were achieved, Earth would not be the sum total of the human habitat. Aided by the rapid gain in technological advances made possible by computerized knowledge, space would be explored, exploited, and settled at a greater speed than might now seem possible, and it will be the space settlements that will represent the new cutting edge of humanity.

Out there on the new frontier, the largest and most nearly unending we have yet seen, risk and danger will be found in plenty. However much Earth will become a quiet center of limited stimulation, there will always remain enormous challenges to try humanity and keep it strong, if not on Earth itself, then on the perpetual frontier of space.

Technology

I have been picturing technology as the prime architect of a livable, and even utopian world of low birthrate. In fact, throughout this book

I have relied on technology as the chief agent for avoiding catastrophe. Nevertheless, there is no denying the fact that technology can also be the cause of catastrophe. A thermonuclear war is the direct product of an advanced technology, and it is an advanced technology that is now consuming our resources and drowning us in pollution.

Even if we solve all the problems that face us today, partly by means of human sanity and partly by menas of technology itself, there is no guarantee that we may not, in the future, be threatened with catastrophe through the continuing success of technology.

For instance, suppose we do develop copious energy without chemical or radiational pollution through nuclear fusion or direct solar energy. Might not this copious energy produce other types of pollution inseparable from itself?

By the first law of thermodynamics, energy does not disappear, but merely changes its form. Two of these forms are light and sound. Since the 1870s, when the electric light was invented by Edison, the Earth's night side has grown brighter by the decade, for instance.

Such "light pollution" is a relatively minor problem (except for astronomers who will, in any case, be transferring the scene of their activities to space before too many decades have passed), but what of sound? The vibration of those moving parts associated with the production or utilization of energy is "noise" and the industrial world is indeed a noisy place. The sound of automobile traffic, of planes taking off, of railroads, of foghorns, of snowmobiles in the winter wilderness, of motor boats on otherwise quiet lakes, of record players, radio, and television subjects us to continuous noise. Will this grow steadily worse and will the world become unbearable?

Not very likely. Many of the sources of unwanted light and sound are under strictly human control and if technology produces them, it can also ameliorate their effects. Electric cars, as an example, would be much quieter than gas-engine cars.

But then we have had light and sound with us always, even in preindustrial times. What about forms of energy peculiar to our own times? What about microwave pollution?

Microwaves, which are radio waves of comparatively short wavelength, were first used in quantity in connection with radar during World War II. Since then, they have not only been used increasingly in multiplying radar installations, but in microwave ovens for fast cooking, since microwaves penetrate the food and turn to heat all through the food instead of, as in ordinary methods of cooking, heating from the outside slowly inward.

But the microwaves penetrate us, too, and are absorbed in our own interior. Could the increasing incidence of stray microwaves in the vicinity of devices that use them eventually have some deleterious effect upon the body at a molecular level?

The danger of microwaves has been exaggerated by some alarmists, but that doesn't mean it is zero. In the future, if Earth's energy supply comes from solar power stations in space, the energy will be beamed from the stations to Earth's surface in the form of microwaves. It will be necessary to proceed with caution in order to make sure that this will not prove disastrous. In all likelihood it won't, but this is not something to take for granted.

Finally, all energy of any kind eventually turns to heat. That is energy's dead end. Earth receives heat, in the absence of human technology, from the sun. The sun is by far the largest source of Earth's heat but minor quantities come from the planetary interior and from the natural radioactivity of the crust.

As long as human beings confine themselves to using the energy of the sun, the planetary interior, and natural radioactivity at no more than the rate at which they are naturally available, then there is no overall effect on the final formation of heat. In other words, we can use sunshine, hydroelectric power, the tides, temperature differences of the ocean, hot springs, winds, and so on, without producing any additional heat over and above what would have been produced without our interference.

If we burn wood, however, we produce heat at a faster rate than would have been produced by its slow decay. If we burn coal and oil, we produce heat where ordinarily none would be produced. If we were to mine for hot water deep underground, we would bring about a leakage of inner heat to the surface at a greater than normal rate.

In all these cases, heat would be added to the environment at a rate greater than it would be in the absence of human technology, and that additional heat would have to be radiated away from Earth at night. To increase the rate of heat radiation, Earth's average temperature would automatically rise above what it would be in the absence of human technology, thus producing "thermal pollution."

To date, all the added energy we have produced, chiefly through the burning of fossil fuel, has not had a very significant effect on Earth's average temperature. Humanity produces 6.6 million megawatts of heat per year as compared with Earth's receipt of 120,000 million megawatts per year from natural sources. We add only 1/18,000 of the total in other words. However, our supply is concentrated in a few relatively restricted areas and the local heating in large cities makes the climate there substantially different from what it would be if the cities were unbroken stretches of vegetation.

What of the future though? Nuclear fission and nuclear fusion add heat to the environment and have the potential of doing so at far higher rates than our present fossil-fuel-burning does. Using solar energy at the surface of the Earth does *not* add heat to the planet, but collecting it in space and beaming it to the Earth does.

At present rates of increase of population and of per capita use, human energy production might increase sixteenfold in the next half-century and it would then represent an amount equal to 1/1000 the total heat production. It might by then begin verging on something that would raise Earth's temperature with disastrous effect, melting the polar ice caps or, worse, initiating a runaway greenhouse effect.

Even if population remains low and steady, the energy we need to carry on a more and more complex and advanced technology will add more and more heat to the Earth, and this might eventually prove dangerous. To avoid the bad effects of thermal pollution, it might well be necessary for human beings to set a sharp maximum to the rate at which energy is utilized—not only on Earth, but on every world, real or artificial, in which human beings dwell and develop a technology. Alternatively, methods might be devised for improving the rate of heat radiation at given, tolerable, temperatures.

Technology can also be dangerous in directions that have nothing to do with energy. We are, for instance, even now, gradually increasing our ability to interfere with the genetic equipment of life, including that of human beings. This is not something entirely new, actually.

As long as human beings have been herding animals and growing plants, they have been deliberately mating them or cross-fertilizing them in such a way as to emphasize those characteristics found useful to men. As a result, cultivated plants and domesticated animals have, in many cases, changed completely from the ancestral organism first made use of by primitive human beings. Horses are larger and faster, cows yield more milk, sheep more wool, hens more eggs. Dogs and pigeons have been bred into dozens of useful or ornamental varieties.

What modern science does, however, is make it possible to juggle inherited characteristics with better aim and greater speed.

In Chapter 11, I discussed how our understanding of genetics and inheritance began and our discoveries as to the intimate role played by DNA.

In the early 1970s, techniques were discovered that would allow individual DNA molecules to be split in specific places by enzyme action. Thereafter they could be recombined. For that matter, a split DNA from one cell or organism could be recombined with another split DNA from another cell or organism even where the two organisms might be of widely different species. By such "recombinant-DNA" techniques, a new gene capable of giving rise to new chemical abilities could be formed. An organism could be deliberately mutated and made to undergo a kind of directed evolution.

Much of the work on recombinant-DNA has been done on bacteria in an attempt, primarily, to discover the intimate chemical details of the process of genetic inheritance. There are obvious practical side effects, however.

Diabetes is a common disease. In diabetics the machinery for manufacturing insulin, a hormone necessary if sugar molecules are to be handled properly within the cells, has broken down. Presumably this is the result of a defective gene. The insulin can be supplied from outside and is obtained from the pancreas of slaughtered animals. There is only one pancreas for each animal and that means insulin is in limited supply and the quantity available cannot be easily increased. Furthermore, insulin obtained from cattle, sheep, or swine is not precisely identical with human insulin.

Suppose, though, that the gene that supervises the formation of insulin is obtained from human cells and added to a bacterium's genetic equipment by the techniques of recombinant-DNA? The bacterium might then be able to form not only insulin, but human insulin, and would pass on this ability to its descendants. Since bacteria can be cultured in almost any quantities, there would become available any reasonable amount of insulin. In 1978, in fact, this feat was performed in the laboratory and bacteria were made to produce human insulin.

Other such feats might be performed. We might design (so to speak) bacteria capable of manufacturing hormones other than insulin; or make them form certain blood factors or antibiotics or vaccines. We might design bacteria that were particularly active in combining the nitrogen of the atmosphere into compounds that would make soil more fertile; or that could perform photosynthesis; or that could turn straw into sugar and waste oil into fat and protein; or that could break down plastics; or that could concentrate traces of useful metals from wastes or from sea water.

And yet what if, quite inadvertently, a bacterium were produced that could cause a disease? It might be a disease to which the human body had never developed defenses, since it would never have been encountered in nature. Such a disease might merely be an uncomfortable one, or a temporarily debilitating one, but it might also be a deadly one, serving as a worse-than-Black-Death laying waste to all humanity.

The chances of such a catastrophe are very small, but the mere thought of it caused a group of scientists working in the field to suggest, in 1974, that special precautions be taken to prevent deliberately mutated microorganisms from getting into the environment.

For a while, it seemed as though technology has given rise to a nightmare even worse than that of nuclear warfare, and pressure arose to end all use of our increasing knowledge of the mechanics of genetics ("genetic engineering").

The fears in this direction seem exaggerated, and, on the whole, the chance of benefits arising from research in genetic engineering are so great, and the chances of disaster are so small and so guarded against, that it would seem a tragedy to give up the former out of a disproportionate fear of the latter.

Still, it will probably be a relief to many if, eventually, those genetic experiments which are seen as risky (along with risky scientific or industrial work in other fields) are conducted in laboratories in orbit about the Earth. The insulating effect of thousands of miles of vacuum interposed between population centers and possible danger would immeasurably reduce the risks.

If genetic engineering, as applied to bacteria, seems to presage a possible catastrophe, what about genetic engineering when applied to human beings directly? This has been raising fears even before the present genetic techniques were developed. For over a century, medicine has been acting to save lives that would otherwise be lost and, in this way, has been lowering the rate of elimination of low-quality genes.

Is this wise? Are we allowing low-quality genes to pile up, and are these serving to deteriorate the human species as a whole until those human beings who are normal or superior can no longer support the increasing weight of defective genes in the species as a whole?

Well, perhaps, though it is difficult to find any way of allowing human beings to suffer or die when they can be easily helped and saved. However sternly individuals might argue in favor of a "hard-nosed" policy in this respect, they might argue less cogently if they themselves or those close to them were involved.

Then, too, the true solution might come with technological advance. The medical treatment of congenital defects is, currently, ameliorative only. Insulin will supply what the diabetic lacks, but the defective gene within the diabetic remains and is passed on to children.* Perhaps the time will come when the techniques of genetic engineering will be used to alter and correct defective genes directly.

Some people fear deterioration of the species through the lowering of the birthrate. The argument is that the birthrate will be lowered to a disproportionate extent by those with better education and higher social responsibility, so that superior individuals will be submerged in a flood of inferior ones.

This fear is accentuated by the claims of some psychologists that intelligence can be inherited. They present data which seem to show that those who are better off economically are more intelligent as well than those who are worse off. In particular, say these psychologists, IQ tests show blacks to achieve consistently lower scores than whites.

The implication is that any attempt to correct what seem to be social injustices are doomed to failure since the oppressed are stupid to exactly the degree that they are oppressed and therefore deserve to be oppressed. A further implication is that population limitation should be practiced more tightly among the poor and oppressed because they are no good anyway.

The English psychologist Cyril Burt (1883–1971), the patron saint of such psychologists, presented data to show that the British upper classes were brighter than the lower classes, that British Gentiles were brighter than British Jews, that British men were brighter than British women, and that British generally were brighter than Irish generally. His data, it now appears, were manufactured by himself in order to demonstrate results that jibed with his prejudices.

Even where the observations would appear to be honest ones, there is considerable doubt that IQ tests measure anything but the similarity of the tested to the tester—with the tester naturally assuming himself to be the cream of the intelligent.

Then, too, throughout history, the lower classes have outbred the upper classes; the peasantry has outbred the middle classes; the oppressed have outbred the oppressors. The result is that almost all the fine superior people of our culture turn out, if we trace back their ancestry, to have come from people who were peasants or otherwise

* A defective gene can arise by mutation in a child with normal parents, so that cruel elimination of individuals will not necessarily eliminate the defective gene anyway.

oppressed and who, in their day, were considered hopelessly subhuman by the upper classes of that day.

It seems reasonable to suppose, then, that since the birthrate must drop, if we are to survive, that we need not worry if the drop does not balance perfectly across all groups and classes. Humanity will survive the shock and probably will not be the less intelligent for it.

Coming closer to our day, a new source of possible deterioration arises out of the scientist's new ability to isolate or produce natural or synthetic drugs that are narcotic, stimulating, or hallucinogenic. More and more otherwise normal individuals seem to be attracted to, and to become dependent upon, these drugs. Will this tendency increase until humanity as a whole has deteriorated past saving?

It might seem, though, that drugs are most valued as methods of escaping boredom or misery. Since it should be the aim of any sensible society to reduce boredom and misery, success in this respect may lessen the danger of the drugs as well. Failure to reduce boredom and misery may produce catastrophe independently of the drugs.

Finally, genetic engineering techniques may serve to guide human change, mutation, and evolution so as to remove some of the dangers we fear. They might actually serve to improve intelligence, remove defective genes, heighten various abilities.

But might not even good intentions go awry? For instance, one of the early victories of genetic engineering might be that of being able to control the sex of children. Might this not radically upset human society? Since it is stereotypical of people to want sons, won't the world's parents choose boys in overwhelming majority?

This is conceivable, and the first result would be a world in which men would greatly outnumber women. This would mean that the birthrate would drop precipitously, since the birthrate depends on the number of women of childbearing age, and only very slightly on the number of men. In an overpopulated world, this could be a good thing, especially since the pro-son prejudice seems to be strongest in the most overpopulated countries.

On the other hand, girls would suddenly gain premium value as competition for them grew keen, and farsighted parents would opt for girls in the next generation as a shrewd investment. In not too long a time, it would be realized that a one-to-one ratio happens to be the only one that really works.

What about "test-tube babies"? In 1978, the newspaper headlines made it appear that one had been born, but that was only a test-tube fertilization, a technique long used for domestic animals. The fertilized egg had to be implanted in a woman's womb and the fetus had to come to term there.

This allows us to envisage a future in which busy career women could contribute egg cells to be fertilized and then implanted in surrogate mothers. Once the baby is born, the surrogate can be paid and the child collected.

Would this be popular? A baby is not, after all, a matter of genes only. A great deal of its development in the fetal stage depends upon the maternal environment; upon the diet of the host-mother, the efficiency of her placenta, the biochemical details of her cells and bloodstream. The biological mother may not feel that the baby she receives from someone else's womb is truly hers, and when flaws and shortcomings (real or imagined) show up in the infant, the biological mother may not patiently and lovingly endure them, but may blame them on the host-mother.

While test-tube fertilization may exist as an added option, it would not be surprising if it proved only minimally popular. We might, of course, go all the way and dispense with the human womb altogether. Once we develop an artificial placenta (no mean task), human egg cells, fertilized in the laboratory, could undergo nine months of further development in laboratory equipment, with aerated nutrient mixtures circulating though them to feed the embryo and carry off the wastes. This would represent a true test-tube baby.

With wombs unused, would women's reproductive machinery degenerate? Would the human species grow dependent on artificial placentas, and be threatened with extinction if the technology failed? Not likely. Evolutionary changes do not take place that quickly. If we used reproductive factories for a hundred generations, the female wombs would still remain functional. Besides, test-tube babies are not likely to be the unvarying route of birth even though they become a possible option. Many women are likely to prefer the natural process of pregnancy and labor if only because they will be more nearly certain that the child is truly theirs. They may also feel that their babies are closer to them for having been nourished by the maternal bloodstream and for having been enclosed by the maternal environment.

There are, on the other hand, advantages to test-tube babies. The developing embryos would be under close observation at all times. Minor faults might be corrected. Embryos with serious deficiencies might be discarded. Some women might prefer the certainty of having healthy babies.

The time may come when we will be able to pinpoint all the genes in human chromosomes and determine their nature. We might be able to locate precisely the seriously defective genes in individuals and estimate the chances of defective children arising from the fortuitous union of defective genes from each of the two particular parents.

Individuals, precisely informed concerning their genetic makeup, may search for mates with genes that would be most suitable for their own, or they might marry for love, but use outside aid for proper gene-combinations in children. By these methods and by outright modification of genes, human evolution might be directed.

Is there the danger here that there will be racist attempts to bring about gene combinations that will produce only tall, blond, blue-eyed children? Or, in reverse, attempts to breed large numbers of dull, moronic people, stolid and patient, to do the world's work and serve in the world's armies?

Both thoughts are rather unsophisticated ones. It is to be assumed that laboratories in many parts of the world will be equipped to do genetic engineering and why should Asians, for instance, dream of producing true Nordic types? As for a race of dull submen—well, in a world without war and with computerized automation, what would there be for them to do?

What about cloning? Might we not bypass ordinary reproduction altogether by taking a body cell from some individual, male or female, and substitute the nucleus of that cell for one in an egg cell? The egg cell could then be stimulated to divide and develop into a baby that would have the precise genetic makeup of the individual who had been cloned.

But why do so? After all, ordinary reproduction is an efficient enough way to produce babies and it has the advantage of shuffling genes to produce new combinations.

Would some people want precisely their own genes to be preserved and given new life? Perhaps, but the clone won't be an exact duplicate. If you were cloned, your clone might have your appearance, but it would not have developed in your mother's womb as you did, and once born, it would have a far different social environment from that which you had. Nor would it prove a probable way of preserving the Einsteins and Beethovens of the future. The clone of a mathematician might not develop mathematical aptitude to a high degree in the clone's own social milieu. The clone of a musician might, under his or her own circumstances, be bored with music, and so on.

In short, many of the fears of genetic engineering, and many of the foreshadowings of catastrophe, are the result of simplistic thinking. On the other hand some of the possible advantages of cloning, for instance, are usually overlooked.

Using genetic engineering techniques not yet developed, a cloned cell might be forced to develop in a distorted fashion so that it would produce a functioning heart with the rest of the body as a vestigial fringe. Or a liver might be thus produced, or a kidney and so on. These could then be used to replace damaged or malfunctioning organs in the body of the original donor of the cell that had been cloned. The body will accept a new organ that is, after all, built up of cells with its own genetic makeup.

Again, cloning might be used to save endangered species of animals. But will evolution, whether guided or not, mean the end of humanity? It might, if we define humanity as *Homo sapiens*. But why must we? If human beings populate space in many artificial settlements that eventually separate and move off into the universe, each on its own, then each will surely evolve somewhat differently and, in a million years, there may be dozens, or hundreds, or a myriad of distinct species, all human-descended, but all different.

So much the better, since variety and diversity can but strengthen the human family of species. We can suppose that intelligence will remain or, most likely, improve, since a species of declining intelligence will not be able to maintain the settlement and will in that way be weeded out. And if intelligence remains and grows, what does it matter if details of outer appearance and inner physical workings change?

Computers

Is it possible that as humanity evolves and, presumably, improves, other species will do the same? Might these other species catch up to us and supplant us?

In a sense, we caught up and surpassed the dolphin, which had a human-sized brain millions of years before humans did. There was, however, no competition between the water-dwelling cetaceans and the land-dwelling primates, and it is only human beings who have developed a technology.

We are not ourselves likely to permit competition; or, if we do, it would be on the basis of allowing another species as intelligent as ourselves to join us as allies in the battle against catastrophe. And, at that, there is no way, unless we encourage the evolution of other species in the direction of intelligence by the use of genetic engineering techniques, for such a catching-up to take place in less than millions of years.

Yet there is another kind of intelligence on Earth, one that has nothing to do with organic life and that is entirely the creation of humanity. That is the computer.

Computing machines capable of solving complicated mathematical problems much faster and much more reliably than humans can (once the computers are properly programmed) were dreamed of as early as 1822. It was in that year that the English mathematician Charles Babbage (1792–1871) began to build a computing machine. He spent years at it and failed, not because his theory was bad, but because he had only mechanical parts to work with and those were simply not sufficiently well adapted for the task.

What was needed was electronics; the manipulation of subatomic particles rather than of gross moving parts. The first large electronic computer was built at the University of Pennsylvania during World War II by John Presper Eckert, Jr. (1919–), and John William Machly (1907–) following a system worked out earlier by the American electrical engineer Vannevar Bush (1890–1974). This electronic computer, ENIAC ("Electronic Numerical Integrater and Computer"), cost three million dollars, contained 19,000 vacuum tubes, weighed 30 tons, took up 1,500 square feet of floor space and used up as much energy as a locomotive. It ceased operations in 1955 and was dismantled in 1957—hopelessly outmoded.

The rickety, unreliable, energy-guzzling vacuum tubes were replaced by solid-state transistors, much smaller, much more reliable, much less energy-consuming. As the years passed, solid-state devices were made smaller yet and still more reliable. Eventually, tiny chips of silicon, a quarter-inch square, as thin as paper, daintily touched with traces of other substances here and there, were made into compact little intricacies fitted with tiny aluminum wires and joined to make microcomputers.

As the 1970s drew to their close, one could obtain for three hundred dollars, from any mail-order house or at almost any corner store, a computer that consumes no more energy than a light bulb, that is small enough to be lifted easily, that can do far more, twenty times faster, and thousands of times as reliably as ENIAC could.

With computers growing more compact, more versatile, and cheaper, they are beginning to invade the home. The 1980s may see them becoming as integral a part of everyday life as the television set became in the 1950s. In fact, earlier in this chapter I referred to the developing computers as the teaching machines of the future. How far will this continue?

So far, the computer is a problem-solving device, strictly bound by its programming, and capable of performing only the simplest operations—but doing so with extraordinary speed and patience. A kind of rudimentary intelligence is beginning to show itself, though, as computers become capable of self-correction and of modification of their programs.

As computers and their "artificial intelligence" take over more and more of the routine mental labors of the world and then, perhaps, the not-so-routine mental labors as well, will the minds of human beings degenerate through lack of use? Will we come to depend on our machines witlessly, and when we no longer have the intelligence to use them properly, will our degenerate species collapse and, with it, civilization?

The same problem and fear must have faced humanity in earlier periods of its history. One can imagine the disdain, for instance, of early builders when the equivalent of the yardstick came into use. Would the cool eye and trained judgment of the skilled architect degenerate forever once any fool could decide what length of wood or stone would fit where by just reading marks off a stick? And surely the bards of old must have been horrified at the invention of writing, of a code of markings that eliminated the need of memory. A child of ten, having learned to read, could then recite *The Iliad,* though he had never seen it before, simply by following the markings. How the mind would degenerate!

Yet the use of inanimate aids to judgment and memory did not destroy judgment and memory. To be sure, it is not easy to find someone nowadays with a memory so trained that he can reel off long epic poems. But who needs that? If our unaided talents no longer demonstrate no longer needed feats, is the gain not worth the loss? Could the Taj Mahal or the Golden Gate Bridge have been built by eye? How many people would know the plays of Shakespeare or the novels of Tolstoy if we had to depend on finding someone who knew them by heart and was willing to recite them to us—if, for that matter, they were likely to have been constructed in the first place without writing?

When the Industrial Revolution brought the power of steam, and then electricity, to the physical tasks of mankind, did human muscles grow flaccid as a result? The feats on the playing field and in the gymnasium belie that. Even the ordinary city-bound office worker can stay in shape by jogging, by tennis, by calisthenics—making up voluntarily for what he need no longer do under the hard grip of enslaved compulsion.

With computers, it might be the same. We would leave to them the rote labors of dry-as-dust calculations, of filing, of retrieving, of record-keeping, thereby allowing us to free our minds for truly creative tasks—so that we might build Taj Mahals in place of mud huts.

That, of course, assumes that computers will never serve for more than the routine and repetitive. What if computers continue to develop without end and follow us to the last stronghold of our minds? What if computers, too, can build Taj Mahals, and write symphonies, and conceive of new, great generalizations in science? What if they learn to mimic every mental ability the human being has? What, in fact, if the computers can be used to act as the brains of robots that will be the artificial analogs of humans, doing all that humans can do, but made of stronger, longer-lasting materials that can better endure harsh environments? Might not humanity become obsolete? Might not the computers "take over"? Might the catastrophe of the fourth class (not merely the fifth) that wipes out human beings be that which leaves behind the heirs they have themselves created?

If we consider this, we might ask a rather cynical question: Why not? The history of the evolution of life is the history of the slow alteration of species, or of the bodily replacement of one species by quite another, whenever that change or replacement results in a better fit within a particular environmental niche. That long, twisting history finally reached *Homo sapiens* a few hundred thousand years ago, but why should that be the final step?

Now that we're here, why should we consider the play to be over? In fact, if we had the capacity to stand back and look at the entire complex path of evolution in world after world, it might seem to us that very slowly, by trial and error, by hit and miss, life evolved until finally a species managed to come into being that was intelligent enough to take the process of evolution into its own directing hands. It might seem to us that only then would evolution really begin to progress as an artificial intelligence, far better than anything that had until then been contrived, came into being.

In that case the replacement of humanity by advanced computers would be a natural phenomenon which, objectively, would be applauded as we ourselves applauded the replacement of the reptiles by the mammals, and to which we could object only out of self-love, with reasons that are essentially frivolous and irrelevant. In fact, if we wish to grow more cynical still, might we not argue that the replacement of humanity is not only not evil, but is a positive good?

I have supposed in recent chapters that humanity would adopt sane steps which would abolish war, limit population, and establish a humane social order—but will it? One would like to think so, but the history of humanity is not exactly encouraging in this respect. What if human beings won't stop their eternal suspicion of, and violence against, one another? What if they can't limit the human population? What if there is no way in which humane decency can be made to direct society? In that case, how can we avoid the destruction of civilization and perhaps even humanity itself?

Perhaps the only salvation lies in the replacement of a species that falls abysmally short by one that will, perhaps, do better. From that standpoint, the fear should be not that humanity will be replaced by computers, but rather that humanity will not be able to advance computers fast enough to prepare heirs ready to take over by the time of the inevitable destruction of civilization.

And yet, what if human beings do solve the problems that face them now, and do launch a decent society based on peace, cooperation, and a wise technological advance in the course of the next century? What if they do this with the invaluable aid of developing computers? Despite human success, might human beings not in any case be supplanted by the things they have created, and would not this be a true catastrophe?

But then, we might ask what we mean by a superior intelligence?

It is entirely too simplistic to compare qualities as though we were measuring lengths with a ruler. We are used to one-dimensional comparisons and understand perfectly what we mean when we say that one length is greater than another, or one mass greater than another, or one duration greater than another. We get into the habit of assuming that all things may be so unsubtly compared.

For instance, a zebra can reach a distant point sooner than a bee can, if both start from the same place at the same time. We are justified then, it would appear, in saying that a zebra is faster than a bee. And yet a bee is far smaller than a zebra and, unlike the zebra, it can fly. Both differences are important in qualifying that "faster."

A bee can fly out of a ditch that holds the zebra helpless; it can fly through the bars of a cage which holds the zebra prisoner. Which is faster now? If A surpasses B in one quality, B may surpass A in another quality. As conditions change, one quality or the other may assume the greater importance.

A human being in an airplane flies more quickly than a bird, but cannot fly as slowly as a bird, and at times slowness may be needed for survival. A human being in a helicopter can fly as slowly as a bird, but not as noiselessly as a bird, and sometimes silence may be needed for survival. In short, survival requires a complex of characteristics, and no species is replaced by another because of a difference in one characteristic only, not even when that characteristic is intelligence.

We see this in human affairs often enough. In the stress of an emergency, it is not necessarily the person with the highest IQ who wins out; it could be the one with the greatest resolution, the greatest strength, the greatest capacity for endurance, the greatest wealth, the greatest influence. Intelligence is important, yes, but it is not all-important.

For that matter, intelligence is not a simply defined quality; it comes in all varieties. The intensely trained and superscholarly professor who is a child in all matters not pertaining to his specialty is a stereotypical figure of modern folklore. We wouldn't be in the least surprised at the spectacle of a shrewd businessman who is intelligent enough to guide a billion-dollar organization with a sure touch, and yet who is incapable of learning to speak grammatically. How then do we compare human intelligence and computer intelligence, and what do we mean by "superior" intelligence?

Right now, the computer can perform mental tricks a human being could not possibly perform, yet that does not cause us to say that a computer is more intelligent than we are. In fact, we are not ready to admit that it is intelligent at all. Remember, too, that the development of intelligence in human beings and in computers took, and is taking, different paths; that it was, and is, driven along by different mechanisms.

The human brain evolved by hit-and-miss, by random mutations, making use of subtle chemical changes, and with a forward drive powered by natural selection and by the need to survive in a particular world of given qualities and dangers. The computer brain is evolving by deliberate design as the result of careful human thought, making use of subtle electrical changes, and with a forward drive powered by technological advance and the need to serve particular human requirements.

It would be very odd if, after taking two such divergent roads, brains and computers would end so similar to one another that one of them could be said to be unequivocally superior in intelligence to the other.

It is much more likely that even when the two are equally intelligent on the whole, the properties of intelligence would be so different in the two that no simple comparison could be made. There would be activities to which computers were better adapted and others to which the human brain was better adapted. This would be particularly true if genetic engineering was deliberately used to improve the human brain in precisely those directions in which the computer is weak. It would, indeed, be desirable to keep both computer and human brain specialized in different directions, since a duplication of abilities would be wasteful and make one or the other unnecessary.

Consequently, the question of replacement need never arise. What we might see, instead, would be symbiosis or complementation; brain and computer working together, each supplying what the other lacks, forming an intelligence-pair that would be far greater than either alone; one that would open new horizons and make it possible to achieve new heights. In fact, the union of brains, human and human-made, might serve as the doorway through which the human being could emerge from its isolated childhood into its in-combination adulthood.

Afterword

Let us look back now on the long trip through the vast wilderness of possible catastrophes that face us.

We might separate all the catastrophes I have described into two groups: (1) those that are probable or even inevitable, like the turning of the sun into a red giant, and (2) those that are extremely unlikely, like the invasion of a huge lump of antimatter making a square hit on the Earth.

There isn't much use in worrying about the catastrophes of the second group. We are not likely to be far wrong if we simply assume they will never happen, and concentrate on those of the first group. Those we can divide into two subgroups: (a) those that loom in the immediate future, like war and starvation, and (b) those that are likely to face us only after anywhere from tens of thousands to billions of years from now, like the warming of the sun or the chilling of an ice age.

Again, there isn't much use in worrying about the catastrophes of the second subgroup right now, since if we don't deal with those of the first subgroup, the rest is academic.

Considering the first subgroup, those catastrophes that are highly probable and that loom close in time, we can again divide them into two sub-subgroups: (i) those that can be avoided, and (ii) those that cannot be avoided.

It seems to me that there are no catastrophes in the second sub-subgroup: there are *no* catastrophes that loom before us which *cannot* be avoided; there is nothing that threatens us with imminent destruction in such a fashion that we are helpless to do something about it. If we behave rationally and humanely; if we concentrate coolly on the problems that face all of humanity, rather than emotionally on such nineteenth century matters as national security and local pride; if we recognize that it is not one's neighbors who are the enemy, but misery, ignorance, and the cold indifference of natural law—then we can solve all the problems that face us. We can deliberately choose to have no catastrophes at all.

And if we do that over the next century, we can spread into space and lose our vulnerabilities. We will no longer be dependent on one planet or one star. And then humanity, or its intelligent descendants and allies, can live on past the end of the Earth, past the end of the sun, past (who knows?) even the end of the universe.

It is that which is, and should be, our goal.

May we gain it!

Index